Language, Pharmacy and Society

AF891331

This book explores the vital role language plays in shaping how we understand and discuss medicines, making for a more detailed study of pharmaceutical and pharmacological language to more clearly understand the intersection of language, health, and culture.

Gonzalez Rodriguez charts the development of the language of pharmacy from the mid-19th century onward, drawing on data from Icelandic and Spanish natural language corpora, historical sources, and contemporary data. The book brings together scholarship from sociolinguistics, media, and cultural studies, and the history of science to highlight the possibilities afforded by an interdisciplinary approach to pharmacy-related language.

The book will benefit readers by providing a deeper understanding of the intersection between language, science, and culture, making it especially valuable for students and scholars in sociolinguistics, history of science, medical humanities, and cultural studies.

Milton Fernando Gonzalez Rodriguez is a Research Fellow at KU Leuven, Belgium.

Routledge Studies in Health Humanities

The Languages of COVID-19
Translational and Multilingual Perspectives on Global Healthcare
Edited by Piotr Blumczynski and Steven Wilson

The Social, Aesthetic, and Medical Implications of Performing Shame
Interdisciplinary Approaches
Marlene Goldman

Pandemic Health and Fitness
Sabina M. Perrino and Joshua O. Reno

Planetary Health Humanities and Pandemics
Edited by Heike Härting and Heather Meek

Language, Pharmacy and Society
The Sounds of Local Knowledges and the Linguistic Logistics of Science
Milton Fernando Gonzalez Rodriguez

For more information about this series, please visit: https://routledge.com/Routledge-Studies-Health-Humanities/book-series/RSH

Language, Pharmacy and Society

The Sounds of Local Knowledges and the Linguistic Logistics of Science

Milton Fernando Gonzalez Rodriguez

Routledge
Taylor & Francis Group
NEW YORK AND LONDON

First published 2025
by Routledge
605 Third Avenue, New York, NY 10158

and by Routledge
4 Park Square, Milton Park, Abingdon, Oxon, OX14 4RN

Routledge is an imprint of the Taylor & Francis Group, an informa business

© 2025 Milton Fernando Gonzalez Rodriguez

The right of Milton Fernando Gonzalez Rodriguez to be identified as author of this work has been asserted in accordance with sections 77 and 78 of the Copyright, Designs and Patents Act 1988.

All rights reserved. No part of this book may be reprinted or reproduced or utilised in any form or by any electronic, mechanical, or other means, now known or hereafter invented, including photocopying and recording, or in any information storage or retrieval system, without permission in writing from the publishers.

Trademark notice: Product or corporate names may be trademarks or registered trademarks, and are used only for identification and explanation without intent to infringe.

ISBN: 9781032989686 (hbk)
ISBN: 9781032989754 (pbk)
ISBN: 9781003601654 (ebk)

DOI: 10.4324/9781003601654

Typeset in Sabon
by Newgen Publishing UK

Contents

Preface

As I noticed soon after I started my research project on the representation of nature and health in West European mass media (1850–2023), understanding the relation between knowledge, health, and culture demands paying special attention to the language of pharmacy. Specially, because it is a sort of language managed, planned and carefully designed to be precise and consistent. One that, for obvious reasons, includes terms not commonly used in other settings. It builds on scientific jargon that includes chemistry- and botany-related lexicon as well as artificially constructed brand names, and linguistic units (e.g., infixes) derived from Latin, Greek, traditional medicinal epistemologies, and century-old local pharmacy practices. Using Icelandic and Spanish as study cases of global and local processes, the book shows that language resembles and can be thought of as a logistic operation. I demonstrate that understanding how medicinal efficacy has been articulated historically demands contextualizing socially the planning, adoption, dissemination, adaptation, and appropriation of pharmacy-related language. Focusing on the language of pharmacy in these two languages, I suggest, allows us to rethink sociolinguistically the place of medicines in our daily conversations.

Pharmacy is a scientific field in the 21st century, but pharmacological knowledge has followed humanity, in one form or another, from times immemorial. Undeniably, local and traditional knowledge—much of which emerged from trial and error—are part of the history of medicine(s). The prefix **-quin-** from Quechua in chloroquine, for instance, is not there by accident. A linguistic logistic process lies behind it. A series of events and actions that emphasize the empirical roots of practices spanning millennia and centuries, and the influence of materia medica exchanges, scientific revolutions, and commercial pursuits. Most, if not all epistemological and ontological systems have safeguarded knowledge and methods on how to utilize plants, animals, and minerals for the care and cure of the human body. Survival has depended on preserving various forms and degrees of pharmacological expertise. Knowing a leaf is curative, or lethal is a matter of life and death, as much as understanding what a pharmaceutical does or can potentially do. Pharmacological knowledge is ancient and has logically left its mark in natural languages. It is therefore important to make a distinction between pharmacy (knowledge) and pharmaceuticals (knowledge-derived objects). Medicines, regardless of their origin or methods behind their creation, are part of knowledge systems as much as linguistic and sociolinguistic ones.

Exchanges of pharmacological knowledge across time and space have left undeniable traces in modern-day speech. Both materially and figuratively, the application of natural precursors has become part of the Western therapeutic landscape. "Exotic" curative plants from faraway places retain their place as a reference in contemporary cultural repertoires, but as I show, in the process, important contributions of medicinal Indigenous expertise to modern science have been overlooked. *Language, Pharmacy and Society* suggests that language can be regarded as a logistic operation, taking place at a major and minor scale. Focusing on the linguistic aspect of medicines identified in corpora, the main question I address in this book is: What does the language of pharmacy tell us about knowledge, science, society, and language itself?

Acknowledgments

I dedicate this book to my parents, María Melba Rodriguez de Gonzalez, and Luis Fernando Gonzalez Parra, as well as to unconditional caring always-there always-knowing-what-to-say Martin, and to my brother Leo, a loyal companion in so many language-related adventures. *Muchas gracias, mange tak*. To Martin, *millioner tak* for sharing his expertise in botany and chemistry.

This book has taken some years to materialize. Along the way I also received the unwavering support of Eric and Irene, to whom I owe so much. The manuscript was originally conceived and first drafted during my time as a visiting scholar at the Department of History and Philosophy of Science at University of Cambridge. I cannot thank Prof. Dániel Margócsy enough for hosting me and providing feedback on the original idea. From the moment the first draft was ready and until the published version was completed, nine anonymous reviewers agreed to read it and to graciously provide their comments. It goes without saying that all mistakes, infelicities, and flaws are my own doing, but I am humbled and will forever be indebted to all those generous scholars who agreed to review the manuscript in its various stages. Their suggestions have been invaluable.

As always, some of the best mentors and true saviors are people we do not know. This book would never have existed without the informants and contributors to the Icelandic and Spanish corpora, as well as all the authors, researchers, support teams and funders behind their creation. My sincere gratitude to all of them. I would also like to express my heartful gratitude to the direction of *Ljósmæðrablaðið* (Midwifery Journal), as well as to Steinunn Blöndal, Rut Guðmundsdóttir, Hólmfríður Garðarsdóttir, Daniela Martín Hidalgo, Liliana Chávez Díaz, Rebeca Fernández Rodríguez, Einar H. Guðmundsson, Ágústa Þorbergsdóttir, Myriam Hélène Blanche Piguet, Aske Hennelund Nielsen, Maritza Gómez Revuelta, Agustí Nieto-Galan, Panagiotis Karagkounis, Austen Van Burns,

Yi-Fan Hu, Alfred Freeborn, Elizabeth Hughes, Cornelius Borck, Robert Meunier, Adriana Churampi Ramirez, Olin Moctezuma-Burns, Silvia Maria Marchiori, Sebestian Kroupa, Franco Orlandi, Luz Pyke, Paula Muhr, Kate Kangaslahti, Flora Lysen, Matilde Carbajo Usano, Brigitte Adriaensen, and Magaly Rodríguez García, who in one way or another, provided encouragement, help, inspiration, challenging fruitful comments, and general advice on the topic of the book.

Some of the chapters have been presented at conferences and events, where engaging audiences and co-panelists were kind enough to offer feedback, useful remarks, and food for thought. I want to especially thank the participants of the Transnational History Lab (Nuremberg), organized by Prof. Maria Rentzetzi, as well as the attendees of the International Congress of the Association of Benelux Hispanists (Leiden), Indigenous Histories of Encounters in Asia-Pacific (Cambridge), Colonial Natures (Cambridge), Biomedical Visions: Aesthetics, Epistemology, and Medical Practice (Berlin), and LOVA Winter School (Amsterdam). A few cases discussed in the book were presented as part of a series of lectures organized by Prof. Kaat Wils (Leuven), to whom I most importantly owe my eternal gratitude for hosting me as a Marie Curie fellow at the Cultural History since 1750 Research Group. This book had never been written without the support of the European Commission, and access to resources provided by KU Leuven, University of Cambridge, University of Iceland, Árnastofnun (The Árni Magnússon Institute for Icelandic Studies), UNAM Library (Mexico City), Ibero-American Institute (Berlin), University of Seville Library, Leibnitz Institute for Educational Media (Braunschweig), Paris Descartes University Library, and Wellcome Library (London).

Many thanks to all of them. In like manner, I would like to extend my heartfelt thanks to Elisse Preposi at Routledge for her support, as well as to Jasmine Erice-Harling, Isobel Brink, Indhumathi Kuppusamy, Janakiram A., and the entire team involved in the production of the monograph.

As for the prehistory behind this book, my first contact with Spanish took place at birth (and obviously some months before that), while I first started learning Icelandic at 11 years of age. Hereby I want to thank all those who have helped me along the way in my passion, curiosity, and interest in these two and many other languages. *Mi sincero agradecimiento. Hugheilar og innilegar þakkir.*

Funding

This book project has received funding from the European Union's Horizon 2020 research and innovation program under the Marie Skłodowska Curie grant agreement No 101026198.

Disclaimer

As a scholar, I am fully aware some of the authors I reference have been involved in actions, behaviors, or ideologies that have caused harm to others. By drawing on their intellectual contributions, I do not in any way endorse these aspects of their affiliations or character, past or present. I remain convinced that meaningful dialogue across time and space and for the benefit of us all begins by engaging with diverse voices and perspectives.

As for technologies used at the preparation stage of this work, AI-based tools, specifically ChatGPT, were utilized in the processing of references.

Abbreviations

Ar.	Arabic
Da.	Danish
En.	English
f.	feminine (gender)
fem.	Feminine
Fr.	French
Gr.	Greek
Ic.	Icelandic
i.o.	indirect object
L.	Latin
lit.	literally
m.	masculine (gender)
M.L.	Medieval Latin
Sp.	Spanish
neg.	negative
pl.	plural
sg.	singular
s.o.	someone
sth.	something

Transcription Symbols

Background noise: <…_...> </…_...>
Defective recording: <defective_record> </defective_record>
Elongation: <elongation/>
Interruption: < />
Language: <… = ""> </language>
Noise: <… .=""/>
Silence: <silence/>
Time: <00:00=""/>

(*) Spelling in Icelandic and Spanish has consistently been preserved as it appears in the original sources.

Introduction

Pharmacy exists to serve, among other purposes, human health and humans discuss pharmacy in their own *terms*. Before this happens though, natural languages acquire a large amount of input from scientific and adopted forms of local knowledge terminology about the substances used to restore health.

Epistemic innovations have a demonstrable impact on language and speech as communities eventually accommodate and respond to new forms of linguistic input. Language will continue bearing the marks of pharmacy in it as much as societies will keep recurring to pharmacy. What needs to be very clear is that pharmacy and pharmaceuticals are not synonyms. Pharmacy is knowledge that communities have accumulated for hundreds, in some cases, thousands of years,[1] based on observation, experience, and some form of trial and error. This knowledge is not a commodity and is an integral part of most epistemic systems around the world, regardless of their understanding of, or relation to scientific practices, in the Western sense of the word. Pharmaceuticals, on the other hand, are materialization of the application of evidence about the pharmacological properties of a substance or preparation. This applicability of pharmaceutical knowledge is (very much)[2] commodifiable and that fact risks perpetuating the unequal access to medicines for those who need them.[3] Remembering that pharmacy is part of the cultural legacy of most living and past societies is crucial to acknowledge that many pharmaceuticals are derivatives of practices embedded in vernacular, traditional and Indigenous local knowledges. As I will show, language has registered some of the history of transfers across epistemic systems, as well as the development of linguistic conventions in sync with scientific innovations and discoveries. From this angle, organizing the production of medicines, I suggest, is not very different than managing logistically the language that embeds the circulation of substances used to preserve and restore human health.

DOI: 10.4324/9781003601654-1

In terms of the choice of linguistic study cases, Spanish is used as an example of a Romance language spoken at the heart of a previously colonizing empire, highly active in the commercialization and circulation of materia medica resources from the Americas for almost four centuries. This allows for the study of remnants of scientific expeditions and mobilization of commodities and artifacts as transformative factors in language (Roldán-Vendrell and Fernández-Domínguez, 2012; Crawford, 2014; Azorín and Sánchez, 2016). Spanish also exemplifies the use of Latin and Greek and the translation from English and French as mechanisms in the production of medicine-related terminology. Spanish is an international language. In contrast, and aside from the case of the West Icelanders who migrated to North America, Icelandic has largely been limited to its use within Iceland. The choice of Icelandic is specifically related to the relatively understudied *nýyrðastefna* (neologisms policy) in pharmacy-related contexts,[4] and the ways in which it reflects the management of language at a social level (Kvaran, 2004a; Kristinsson, 2018). Icelandic provides insights into the creation and adoption of medicinal technicisms constructed, for instance, through revitalization of ancient words and through photo- semantic calques. From a sociolinguistic perspective, the case of this Nordic language is also particularly interesting due to the involvement of various actors in the creation, introduction, and adoption of scientific terms.[5] Whereas medicinal lexicon in Spanish tends to obscure the etymological origin of technicisms, and by proxy their meaning, Icelandic neologisms distinguish themselves for their self- explanatory nature. Throughout the entire book, examples in Icelandic and Spanish alternate to render clear in which ways the language of pharmacy is equally present in both languages, couched in two different linguistic systems, both equally related to traces of traditional knowledge, novelties in botany, advances in medicine and chemistry, increased understanding of pharmacology, marketing of medicines and commercial strategies.

We discuss the substances we use to preserve and restore our health and that means that medicines have a biosocial,[6] as well as a sociolinguistic component to them.[7] The intersection of language and medicines is important because of the presence of pharmacy in the lives of patients, caregivers, and communities. After all, thousands of medicines are developed,[8] produced and consumed every day and patients spend trillions of dollars in procuring pharmaceuticals.[9] This has linguistic implications.[10] Despite the complexity of pharmacy- related concepts and names of medicines, they routinely find their way into daily conversations. They enter our bodies as much as they enter our therapeutic imaginaries, but also linguistic repertoires. A special type of vocabulary is used when we talk about pharmaceuticals and remedies. The language of pharmacy has its own sounds. But multisyllabic, slightly unpronounceable names are not only references to technoscientific objects designed to keep us healthy. They are also an integral part of our lived experience. Probably everyone on Earth has at some point taken a pill, received an injection, purchased an over-the-counter tablet, or used a plant-, mineral- or animal-based substance for health management purposes.

Science depends on the presentability and representability of knowledge to allow for meaningful exchanges that secure its progression. In this sense, it is no different than any other type of knowledge. It depends on a facilitating mechanism that makes transmission possible, and if feasible, through transfers embedded in a type of communication. Here is where language, knowledge, and science fuse.[11] Elements in pharmacy can derive from marks and markers of trial-and-error episodes occurring on a laboratory table or in the setting of a healing intervention in a non-urbanized context. These episodes that we could call epistemic episodes for the sake of the argument are also sites of production of object-depending knowledge. The origins of pharmacy can be traced back to the processes of discovering and identifying medicinal substances. Since this is the case across all versions of knowledge-based therapeutic treatments, even in non-scientific settings (according to Western definitions of science), all languages inherently have words to express the concept of "medicine", "drug", "healing object", "healing precursor", "potion", "remedy", etc. The language of (bio)medicine has accommodated the vocabularies to discuss approaches to healing since times memorial.[12]

In the context of Western (bio)medicine, which is the departing point of my argument, history has recorded the onset, development, and evolution of pharmacy as a discipline, field, and ground for life-saving knowledge production.[13] In parallel, pharmaceuticals as end-products lie at the heart of commercial enterprises, networks, and structures, in many ways, linked to profit-driven medicalization initiatives at a large scale and one of the raison d'être of the enormous economic gains the commercialization of medicines generates.[14] Pharmacy and commerce are intertwined. Developing and producing medicines demands high amounts of capital. Even in the context of state-run healthcare systems, the costs of pharmaceuticals must be covered *somehow* by *someone* along the line. As I will show in the subsequent chapters, the commercialization of medicines is a major source of linguistic production. Words to refer to pharmaceuticals derive from highly scientific settings, in the case of concepts (e.g., "vitamin"), but also from highly marketing-driven circuits (e.g., Viagra).[15] Much is at stake as these are multimillion investments (i.e., top-notch facilities, rigorous procedures, expensive raw materials). Even more is at stake if one considers the health of billions of humans and the precarious situation of those who cannot have access to medicines due to their high costs. Pharmacy, without any doubt, is one of the most dramatic examples of the mercantilization and materialization of knowledge;[16] a double-edged sword and relatively recent development. Medicines are needed to preserve and restore health, but many are left behind the paywall. The connection between medicines, commercial interests, and language is obvious. Potential buyers, especially in the case of over-the-counter products, need to presumably know, understand, and proactively want to administer themselves or others' medicines.[17]

Empires and Colonies: Linguistic Cores and Peripheries

Whereas Icelandic and Spanish have had quite opposite approaches to the adoption of specialized terminologies (e.g., stable consistent purism versus moderate discontinuous purism), these languages share certain important similarities (e.g., adoption of scientific terminology through translation). In both instances, language acts as a mediator in the careful articulation and distribution of knowledge hierarchized by external factors and adopted linguistically in sync with standards determined elsewhere. Icelandic and Spanish illustrate how linguistic communities outside the primary centers of scientific knowledge production in the 19th and 20th centuries (e.g., Germany and the United States) have adapted and responded to advances in science.[18] Spanish, as a global language spoken across three continents by hundreds of millions, contrasts with Icelandic, which is spoken by a small, localized community on an island nation. Both languages provide a case study for understanding the local versus global dimensions of pharmaceutical language historically, offering insights into how different sociolinguistic communities incorporate scientific terminology into everyday language in the 21st century.

Linguistic planning in Iceland developed alongside the establishment of various academic fields in the 18th to early 20th centuries. The introduction of scientific discoveries from Europe and North America was closely tied to print media, especially translations. These translations of key academic works and textbooks exposed Icelandic readers to a significant influx of new scientific terms, concepts, and neologisms. Icelandic scholars educated abroad led efforts to adapt (Icelandize) these volumes, despite challenges such as the absence of established technical terminology and limited support from Danish authorities.[19] This process was crucial for introducing subjects like chemistry, a precursor to pharmacy, to Icelandic readers. As discussed in Chapter 2, the development and eventual standardization of the term "efnafræði" for chemistry, for instance, followed a similar trajectory to many words integrated into Icelandic during the long 19th and early 20th centuries. Key individuals, often educated in Denmark played an instrumental role in creating and promoting terminology that captured the latest scientific advancements in Europe. The role of these intellectuals in shaping Icelandic neologisms in response to pharmaceutical advances and novelties is comparable to the influence of Mexican physicians in shaping pharmacognosy terminology.

Entries in Indigenous languages such as Nahuatl, Maya, and Otomi in 19th- and early 20th-century Mexican pharmacopoeias highlight the intersection of scientific, colonial, political, and commercial processes. These entries reveal the crucial role of linguistic planning in the transfer and dissemination of medicinal knowledge, as cataloging botanical specimens required the development of terminology to describe them. Although sometimes regarded as colonial tools, pharmacopoeias provide insights into the hidden sources of knowledge, while the gradual disappearance of Indigenous terms reflects the evolving dynamics of pharmacy-related vocabulary. They illustrate the hierarchical relationship between global and local knowledge, offering a core-periphery view of pharmacological history. The knowledge of pharmacological properties in nature has been preserved through generations in many societies, predominantly through linguistic devices (e.g., oral tradition). It is hard to imagine conveying information about medicines without a certain level of detail and verbal communication. Even today, the use of natural remedies is closely tied to oral traditions in various parts of the world. The acquisition of specimens spurred the movement of plants, animals, and minerals across Western labs, museums, universities, and factories, leaving linguistic imprints not only on labels and catalogs but also in pharmacognosy textbooks, training materials, administrative records, and advertisements. These sources, though not always intentionally, provide invaluable insights into the language of pharmacy during their time. From a colonization-of-knowledge perspective, these documents, while reflecting European dominance, can also be viewed as unintentional monuments to the epistemic legacy and linguistic heritage of marginalized voices suppressed through time and space.

In the realm of pharmacological advancements, Iceland, Latin America, and Spain were largely on the receiving end, rather than the pioneering side, for much of the 19th and early 20th centuries. However, language planning policies for adopting biomedical terminology have varied significantly between Icelandic and Spanish. Iceland's implementation of its *nýyrðastefna* (neologism policy) has unique characteristics. While Spanish has incorporated numerous terms from French in the 19th century, from English and German in the 20th century, and predominantly from English in the 21st century,[20] linguistic purism has only been partially promoted and inconsistently applied across the Spanish-speaking world.

Pairing Icelandic and Spanish as case studies involves comparing the linguistic and epistemic systems that shaped the development and implementation of modern pharmaceutical language. The evolution of pharmacy-related terminology in both languages has followed distinctly different paths across various historical periods. While Spanish has been shaped by global exchanges, external influences on both sides of the Atlantic, and the integration of Indigenous terms, Icelandic has gone through periods of Danish influence, linguistic purity, and coordinated deliberate efforts to keep pace with science and technology. The fact that both languages have experienced varying degrees of diglossia in relation to Latin, Greek, English, German, French, and Danish presents new viewpoints on how scientific knowledge is sociolinguistically shaped within different cultural contexts.[21] Due to the miscellaneous nature of pharmacy, the commercial aspect is intrinsically linked to advertisement (i.e., physicians, druggists, and patients need to be informed about new drugs). The unique histories of the emergence and expansion of mass media outlets offer fresh perspectives on the development of pharmaceutical collective imaginaries within two distinct linguistic communities. As the upcoming chapters will show, a comparison of the way in which scientific language has been presented across advertisement in the two languages, provides new insights about the interplay between science, communication, and linguistic standards. In similar ways, the Spanish-Icelandic comparison allows for an exploration of sociolinguistic prestige and the role of language in establishing scientific authority. Global exchanges of knowledge and materia medica led to the incorporation of local medicinal knowledge from Latin America into significant pharmaceutical advancements (e.g., quinine). Language played a crucial role in establishing the authority of this knowledge and facilitating its global dissemination. In contrast, Iceland has historically relied on the development of biomedicine outside the country. Both Spanish and Icelandic languages retain traces of these distinct histories. This comparison provides new insights into how each sociolinguistic community has adopted, adapted, and appropriated terminology related to medicines, ethnomedicines, and pharmaceuticals, and deploys it in everyday language.

As discussed in Chapter 2, Spanish became an important vehicle to mobilize large amounts of medicinal knowledge for over four centuries. The role, scale, and impact of the Spanish Empire on the exchanges of materia medica between Europe and the Americas, Africa, and Asia-Pacific shaped the scientific landscape irrevocably. A comparison between the case of Icelandic and Spanish, offers insights about global versus local processes, but also about the contexts of relative isolation versus transcontinental contact. Through various colonizing projects, Spanish as an imperial language, operated as a linguistic broker and facilitator of epistemic pharmacological transfers. For instance, through the composition or translation of herbals and medical literature, such as the Codex Badianus in the 16th century (Figure 0.1).

The ethical and political aspects thereof are questionable, nonetheless because the "encounters" between local communities and the Empire

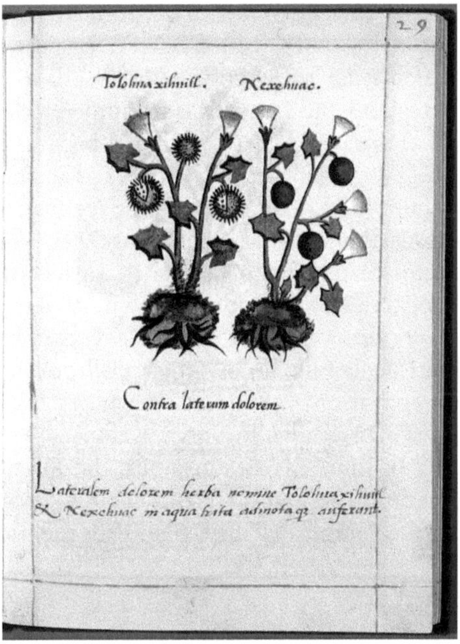

Figure 0.1 Badianus Codex, (facsimile), 16th century: species of Datura, written in Nahuatl by the Aztec physician Martinus de la Cruz, translated by Juan Badiano into Latin, presented to the son of the first Viceroy of New Spain in 1552. Two species of Datura used for "pain in the side". Courtesy of Wellcome Images.

were not always pacific, mutually beneficial, agreed upon, or equilibrated (Garðarsdóttir, 2023). The incursion of new knowledge brought with it new sounds. Many "encounters", or discoveries, depending on how one looks at them, materialized in an expanded repertoire of (re)sources, many of which became staples in pharmacists' shelves across Europe. Indigenous materia medica and the names that came with them enriched Western catalogs of medicines. Spanish, along with the languages spoken by other major colonizing superpowers, such as Dutch, English, and French, had an impact on the *terms*, literally and figuratively, on which the adoption of materia medica took place.

A tradition of preserving botanical and animal- or mineral-based preparations mentioned in important pharmacopoeias or previous editions probably explained the inclusion of substances difficult to obtain locally.[22] An example thereof is "Liquen de Islandia" (*Centraria islandica*)[23] [Icelandic moss], mentioned on page 108 of *Nueva Farmacopea Mexicana de la Sociedad Farmacéutica de México* (*New Mexican Pharmacopoeia of the Pharmaceutical Society of Mexico*) from 1896.[24] On the lichen, Hooper, and Grant write that "it had once a very high reputation for curing phthisis, which it has now entirely lost" (1386). The knowledge about the lichen in Iceland and Lapland is presumed to have been kept for centuries, and as early as 1673, it would have been available in Danish pharmacies.[25] In the 18th century, it was included in the list of "World's Commercial Products" (1725) for being "an excellent food for invalids, and medicinally it is used in cases of lung disease" (p. 85). The same properties as the Mexican pharmacopoeia report. In Iceland, the medicinal application of the local moss was widely known. As the entry (Figure 0.2) in *Um íslenzkar drykkurtir, söfnun þeirra, geymslu, nytsemi, verkanir og tilreiðslu* (About Icelandic herbs, their collection, storage, usefulness, effects, and preparation) (1860) by Alexander "Farmer" Bjarnason illustrates.

Best collected dry in June, Bjarnason notes, the lichen has nourishing, blood- purifying, vermifuge, and strengthening properties. Building on medicinal literature published in Iceland at the time, he reiterates

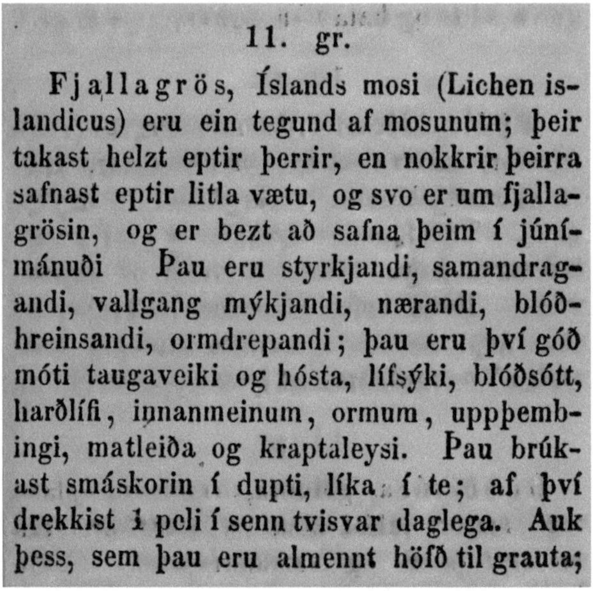

11. gr.

Fjallagrös, Íslands mosi (Lichen islandicus) eru ein tegund af mosunum; þeir takast helzt eptir þerrir, en nokkrir þeirra safnast eptir litla vætu, og svo er um fjallagrösin, og er bezt að safna þeim í júnímánuði Þau eru styrkjandi, samandragandi, vallgang mýkjandi, nærandi, blóðhreinsandi, ormdrepandi; þau eru því góð móti taugaveiki og hósta, lífsýki, blóðsótt, harðlífi, innanmeinum, ormum, uppþembingi, matleiða og kraptaleysi. Þau brúkast smáskorin í dupti, líka í te; af því drekkist 1 peli í senn tvisvar daglega. Auk þess, sem þau eru almennt höfð til grauta;

Figure 0.2 Entry for Centraria islandica (basonym: *Lichen islandicus*) in Um íslenzkar drykkurtir, söfnun þeirra, geymslu, nytsemi, verkanir og tilreiðslu, by Alexander "Farmer" Bjarnason (1860), (page 39, page 35 in the original).

the therapeutic applications of *Centraria islandica* (basonym: *Lichen islandicus*) in the treatment of dysentery, indigestion, whooping cough, and lack of energy, among other conditions. Either, "smáskorin í dupti, líka í te" (chopped into powder, but also in tea form) (Bjarnason, 1860, p. 35), it has many therapeutic applications. With few objections, Mitchell (2015) notes that "Linnaeus' endorsement ensured the continued use" (59) of lichen-based remedies as part of 18th-century materia medica. Despite a considerable number of entries on lichens in medical compendia, after several pitfalls, lack of confidence and the emergence of novel therapeutic substitutes, *Centraria islandica* "had become the sole lichen of any medical interest, and so it remained for almost a century" (Mitchell, 2015, p. 63).[26] The lichen keeps appearing in Spanish-speaking medical literature throughout the 19th century and early 20th century. Importantly, although Icelandic lichen did not grow in Mexico, a replacement could be imported from several places in North and West Europe where moss grew in the coldest areas.[27] Locally, it was retained in the therapeutic repertoire and is currently undergoing a period of revitalization.[28] Health plays a vital role in the relationship between knowledge and society, providing a valuable lens through which to examine the function of language in this dynamic, both socially and historically. Much remains unknown on the sociolinguistic aspect of pharmacy, nonetheless within small linguistic communities, such as the case of Icelandic.

Drugs constitute an important element in the relation and dependencies between societies and knowledge, offering an entry point to study the role of language. In the 21st century, there are several examples of lexical traces of traditional knowledge in the language of pharmacy. Although it might seem that Indigenous pharmacological knowledge has not always left traceable marks in the history of pharmacy, according to the World Health Organization, in the 21st century, "(o)ver 40% of pharmaceutical formulations are based on natural products and landmark drugs, including aspirin and artemisinin, originated from traditional medicine" (2024, n.p.). Many have been the encounters between localized forms of medicinal knowledge and science. Some of the terms preserved, and which derive from Indigenous knowledge systems include kavatin (from *kava*) from Fijian and Maruesan,[29] laurepukine (from *pukatea*) and tutin (from *tutu*) from Māori, protokosin (from ኮሶ, *kosso*) from Oromo, cherimoline (from *chirimuya*), guanine (from *wanu*) and ratanhine (from *ratánya*) from Quechua, boldine (from *folo*) from Mapudungun, among others (Senning, 2019). Other cases are less evident. Introcostrin, for instance, stemmed from a form of curare,[30] discussed in Chapter 3, void of any lexicographic, phonetic, or morphological reference in its name.[31] Scientific terminology retains elements of the linguistic heritage of several Indigenous languages, as well of linguistic planning policies.

In the context of the language of pharmacy, defined by its ancestry anchored in chemical, medical, and botanical terminologies, there are other scenarios that resemble the case of Spanish. Arabic, French, Chinese, and Portuguese, for instance, have retained markers of adoptions of materia medica from other linguistic geospheres. Arabic and French serve as a basis for several chemical terms in the international language of chemistry.[32] Some of them include civetic acid (from زَبَاد, zabād), curcumene (from كُرْكُم, kurkum), safranal (from زَعْفَرَان, zaʕfarān) from Arabic and caffeine (from *café*), ergoline (from *ergot*), gelatin (from *gelatine*), and urea (from *urée*) from French (Senning, 2019). English maintains dominance as a "hegemonic language" (Gordin, 2015, p. 2017) in the field of health, and by extension, pharmacy; even in the case of speakers of languages with a long-standing tradition of scientific thought and knowledge dissemination. Anglicization has been adopted, accommodated, and negotiated differently,[33] for instance, in Arabic,[34] Persian,[35] Russian,[36] and Turkish.[37] So far, the interplay of language, interlinguistic pharmacological knowledge dissemination and society has been studied in the context of epistemic and linguistic encounters across West Africa,[38] East Asia,[39] India,[40] and the former Soviet Union geosphere.[41] Preventing, diagnosing, treating, and eradicating diseases lies at the core of any survival project for every single society around the planet. Hence, all languages presumably have ways to communicate, describe and discuss health-related practices and objects. Remembering again that pharmacy and pharmaceuticals are not synonyms. In the Western context, medicinal, pharmaceutical, and therapeutic knowledge is expressed linguistically in sync with the advances, novelties, and discoveries set in motion by empiric research.[42] This explains why the biomedical language has exponentially grown in the 20th and 21st centuries and it will probably continue to do so. After all, it is mainly through language, along with other formats of information transmission, that societies learn and pass on large bodies of knowledge about medicines.

Icelandic and Spanish: Global and Local Sounds

It is through a specific type of language that medicines are marketed and presented to the consumers. For whom, a proper noun denoting an object with pharmacological properties becomes part of their linguistic repertoire, acting as referents while also acquiring and activating semantic and pragmatic references.[43] One example from the corpora is the introduction and sociolinguistic adoption of the term "imodium" in Spanish, as direct result of the commercialization of loperamide hydrochloride ($C_{29}H_{33}ClN_2O_2$) as a prescription (early 1980s) and over-the-counter antidiarrheal (late 1980s), under the brand names Imodium®,

Imodiumcaps® and Imodium A-D®.[44] By the early 1990s, *imodium* had established itself as a pharmaceutical, but also as a linguistic addition, in the parts of the Spanish-speaking world where it was advertised and sold. While language and society adopted Imodium®, they also adapted *imodium* through a linguistic negotiation process. Imodium®/*imodium* has been semantically and pragmatically hispanized. This is best illustrated through the following example taken from the Sociolinguistic NOW Spanish Corpus in the context of a violent crime that takes place in Texas:

1 En Texas existe pena capital asi que en este momento estan pidiendo imodium por cajas.
[*They have death penalty in Texas so they are ordering boxes and boxes of Imodium*][45]

In this example, a Puerto Rican speaker uses "imodium" to refer figuratively to the state of fear he presumes the perpetrators of the crime must be experiencing. According to him, it would be such that they would require large quantities of antidiarrheal medication. In this specific context, language provides evidence of the traces of loperamide hydrochloride's history,[46] subsequent commercialization, and circulation epistemically (pharmacological efficacy), but also linguistically (synonymy > *antidiarrheal*).

Exchanges from the corpora also reveal the interwoven presence of drugs in the household therapeutic repertoire. Familiarity with the medicine is expressed through the colloquial register. In the next example from an Icelandic corpus, the feminine form refers presumably to "töflur" (*pills*), which from context, are not assumed to matter. As it is the case in various examples identified in the corpus, the drug ibuprofen is assigned either no grammatical gender, or a partitive defined by the means of administration. Tablets in this case is a feminine noun.

2 En spurning mín er sú að hann var svo mikið að drepast að ég gaf honum paratabs áður en við fótum til læknis og virkaði ekkert, svo eftir lækninn gaf ég honum eina íbufen og ekkert virkaði og hann var bara að kálast þannig ég bætti smá panodil zapp við og enn gerðist ekkert og ég bað hann alltaf að bíða og gefa séns og borða (já eða drykkjarjógúrt og mauk því hann getur varla hreyt kjálkann) og þegar ég kom heim setti ég eyrnadropa frá jurtaapótekinu í eyrað og allir verkir snarhættu, haldið þið að verkjatöflurnar hafi allt í einu kikkað inn eða var þetta svona svakalega góð virkni dropanna [sic] (2014).
[*But my question is about the fact that because he was so much in pain, I gave him paratabs before we went to the doctor and nothing*

> *worked, then after the doctor I gave him one ibuprofen and nothing worked and he was suffering so much that I added some panodil zapp and still nothing happened and I kept asking him to wait and give it a chance and try to eat (right or drinking yogurt and puree because he can hardly move his jaw) and when I got home I put ear drops from the herbal pharmacy in his ear and all the pain immediately went away, do you think that the painkillers suddenly kicked in or was it just that the drops worked so marvelously well*][47]

This excerpt from the MÍM (Mörkuð íslensk málheild) (Tagged Icelandic Corpus) database provides insights about the sociolinguistic of pharmacy from various vantage points. Firstly, from a biosocial and intimate perspective, the caregiver experiences great amounts of frustration about the ineffectiveness of the medication, but also the assistance provided by the physician. The specifics about the type of Panodil (acetaminophen) administered by adding the full name provides additional information on the attempt to manage the pain as effectively as possible. Both trade names of paracetamol, Paratabs, and Panodil Zapp are cited and mentioned as part of the overall care the patient received. Taking into consideration the full text, it suggests that a certified doctor did not oversee the entire selection of medicines administered to the person in question. Secondly, from an epistemic perspective, the passage also reveals some of the implications of agency in the context of medicines and hierarchies of knowledge production. The "jurtaapótek" (botanical pharmacy) mentioned in the text presumably sells over-the-counter "eyrnadropa" (eardrops) the author of the post does not mention by trade name. Most likely, she/he is aware that whereas the pharmaceutical vocabularies of the readers would include "paratabs" and "panodil zapp", the brand name of the botanical preparation might not *sound* familiar to the target audience. The most plausible reason is because pharmaceuticals are linguistically more present in Icelandic than medicines produced outside biomedicine-related protocols or canons. Thirdly, the category "verkjatöflur" (*painkillers*) is used to categorize the two previously mentioned medications, but excluding the ear drops, despite their presumed pain-soothing effect. The final sentence about the "góð virkni dropanna" (*good efficacy of the drops*) encapsulates one of the main reasons that drives the relation between health and societies: medicinal efficacy. More importantly, this entry in the corpus offers also insights about the place of pharmaceuticals vis-à-vis non-pharmaceutical preparations in the care and cure of the human body. Science is partially questioned, or at best, compared with a plant-based formula in its capacity to relieve and manage pain. No brand names are mentioned to describe more specifically which botanical products were used. Neither is the source of expertise added to contextualize how the eardrops were identified, or to credit the staff at the botanical pharmacy for their advice or expertise.

The inclusion of Paratabs, Parkodin Zapp, and an undisclosed brand of herbal eardrops suggests that linguistic marketing plays a fundamental role in positioning a medicine in the imaginary of potential patients.[48] Brand names come to refer to the set of medicinal properties and anticipated benefits expected from them. Their biological effect on the body, as much as their pharmacokinetics and pharmacodynamics are reduced to catchy, memorable words. As in the above-mentioned example, Paratabs is not described, its name is self-explanatory. The name of an anonymous vessel of botanical eardrops would have not delivered the message. One might speculate that even the name of the botanical pharmacy might have also failed to provide context. What is it about the names of pharmaceuticals that make them part of our collective memory and facilitate their incursion in the linguistic repertoire of millions? A fair question considering the atypicality of long names that often do not phonetically or morphologically align with the language spoken by the patients. Howard Ansel (1965) observed that, besides pronunciation, the secrecy behind certain brand names risked confusing physicians, druggists, and patients. For Ansel, the pursuit for originality already in the 1960s complexified and obscured the capacity of users to work out from a brand name what a medicine did. In the recent examples cited, as a name, Paratabs offers indeed few cues. So do five-syllable "i-bu-pro-fe-no" (*Sp.*) and trisyllabic "í-bú-fen" (*Ic.*). What is clear in both instances is that these are pharmaceuticals. By proxy, they are products of technological processes guided by the instrumentalization of science and of knowledge about the human body. In the previous excerpt, a distinction is made between (bio)medicine and ethnobotany as epistemic practices. A difference in perception, approach, and epistemic hierarchy gleans from the language used to refer to technoscientific products and traditional knowledge artifacts.

Science, Commerce and Media

Language is instrumental in conveying changes and developments in knowledge systems, particularly in linguistic communities which have predominantly been on the receiving end of major scientific revolutions. This has been the case for Icelandic and Spanish. Reviewing the planning, management, and circulation of pharmacy-related terms across time, I suggest, allows us to understand better the role of science in daily communication. A comparison of pharmaceutical language in two different languages offers meaningful insights about the role of science, commerce, and media in the sociolinguistic adoption of medicines. Firstly, because providing information about medicines across mass media, among other things, involves describing ingredients, properties, and therapeutic effects. Many of which require and depend on the adopted localized standards for translation and adoption of terminology. Secondly, because validating claims about the medicinal efficacy of a product or brand requires often the use of technical and specific terms to explain the science or medicinal knowledge behind their creation.[49] Media outlets mobilize and alert about novelties, examples of which are provided in Chapters 3, 4, and 5, for Iceland, Latin America, and Spain. Thirdly, because discoveries about human health and sickness are in constant evolution, prompting a continuous revision of terminologies that reflect additions or changes in knowledge systems and in one of their material by-products: pharmaceuticals.

Throughout the 20th and 21st centuries, audio- visual and written sources have registered the influence of pharmacological knowledge and its linguistic traces in daily communication. Today's media outlets clearly show the increased presence of linguistic elements from fields related to biomedicine. A type of knowledge that demands scientific rigor and standardization. Medical, and hence pharmaceutical language, has put to the test the plasticity of languages to keep pace with the influx and implications of scientific terminology. This has had sociolinguistic and mediatic repercussions, such as the case of the "píldora" (*pill*) in the Spanish-speaking world; discussed in Chapter 5. Aside from major biosocial revolutions, such as contraceptives, the history of the language of pharmacy gleans interesting insights about the society- science dyad. Examining historical excerpts from Icelandic and Spanish pharmaceutical advertising, as is done in Chapters 2, 3 and 4, offers fresh perspectives on how science has been communicated across various contexts and scales.

Vioftalyl, a formula used to treat amoebiasis, is for instance one of those pharmaceutical names that have left minimal linguistic traces in Spanish. As the leaflet targeted at specialists reads, Vioftalyl "proporciona resultados óptimos en el tratamiento de las infecciones intestinales amibianas o banales, gracias al efecto combinado y sinérgico de sus componentes: ftalil- sulfatiazol, diyodohidroxiquinoleina and pectina citrica" [*provides optimal results in the treatment of amoebic or banal intestinal infections, thanks to the combined and synergistic effect of its components: phthalyl- sulfathiazole, diiodohydroxyquinoline and citrus pectin*] (Bigaux, ca. 1960, n.p.).[50] It was produced locally by a laboratory,[51] André Bigaux,[52] that had previously manufactured coal tar and other chemicals.[53] Different from the publicity display Bigaux allocated during the 1970s, for instance, to Viritab,[54] Vioftalyl was mainly advertised via physicians. Or as part of announcements in catalogs and newspapers, such as the limited-time of a 60-ml jar of Vioftalyl for 2,204 MXP (today MXN), as opposed to 2,896 MXP, posted in *El informador* on 14 June

1989 (p. 2F). A special *deal* offered by the Levy pharmacists' chain on the occasions of the "Día de antidiarreicos y antisépticos intestinales" [*Day of antidiarrheals and intestinal antiseptics*], suggestively advertised as the perfect opportunity to buy medicines for prices that caused "verdadero alivio" [*real relief*] (p. 2F).

But as Unda (1986) observes in his report on the abuse of antidiarrheal agents and antimicrobial drugs in the treatment of infant diarrhea in Mexico: "It is evident from the brochures and pamphlets that there is little control or restriction on the marketing or publicity of these products" (p. 5). The language was opaque and yet, the images and instructional material indicated the formulas were safe for adults and children. His report suggests a tendency to overconsume medicines for conditions where diarrhea is one of the symptoms, nonetheless because pharmacists dispense these medicines without a doctor's prescription. In his survey of 12 pharmacies in Mexico City and vicinity, Unda found out that often the problem lay in the fact that pharmacists consulted *Diccionario de Especialidades Farmaceuticas* (Physician's Desk Reference), following taxonomical entries, such as "antidiarrheal", but without delving into the ingredients of each formula. Similarly, the public was (and is) exposed to advertisements containing terminology unknown to them,[55] except for specific descriptors that disguise crucial information about the medicines they consume.[56] Building in the concept of "semiotic assemblages" (Pennycook, 2021, 2024) to approach advertisement of pharmaceuticals, it is possible to consider them as examples of dynamic inventories and reconfigured repertories that build on language and images to deploy a message. Vioftalyl, for instance, depends on the accompanying texts to explain its amoebicide properties. The linguistic nature of the captions of the images explains the therapeutic identity of the object being advertised. It is by means of the semiotic conventions understood by physicians (i.e., the target audience) that the medicines enter their mental catalog of options for their patients.

The encounter of audiences with advertisement might also be considered in terms of semiotic encounters and semiotic precarity, as defined by Lee (2022). Whereas the first is defined as "a moment in which a given semiotic resource performs meaning (signs, after all, have no meaning unto themselves but acquire meaning when they are intersubjectively activated)" and the second as "as a condition imminent in spaces in which the quotidian and unremarkable specificities of cultural difference are unpresumed (either uncertain or questionable) and, as a result, come to be both remarkable (in the sense of literally being worthy of remark) and as semiotically distinct" (p. 21). From this angle, much of the terminology used in the promotional material destined for doctors is semiotically precarious to the readers not familiar with it and yet, it remains meaningful. It looks scientific and hence, convincing. A similar phenomenon to the phonetic notation of

Indigenous names of plants discussed in Chapter 2 and which can also be seen as examples of "pictograph-ness", "the qualities of visual surfaces to merge texts and images as pictorial strategies" (Gonzalez Rodriguez, 2022a, p. xxvi). Linguistic strategies in the branding and promotion of medicines can potentially conceal and deceive unguarded consumers and patients, but also serve a purpose in tactically outmaneuvering obstacles in the free access of pharmaceuticals. Language moderates science communication, but the semiotics of terminology, simultaneously visual and linguistic, play a decisive role in the dissemination of information about medications across media. Using examples from pharma-marketing in Icelandic and Spanish, some of the upcoming chapters explore the impact of commerce and media on the sociolinguistics of pharmacy.

Theoretical, Methodological and Conceptual Considerations

This is primarily a medical humanities monograph, written from an interdisciplinary perspective. In terms of the theoretical and methodological approach, the book consistently adheres to the empirical data gathered from corpora. The focus and starting points are the sociolinguistic evidence related to the language of pharmacy in Icelandic and Spanish. Throughout the examination of this data, historical sources and various philosophical and anthropological (STS) works are extensively utilized. However, the study maintains its primary identity as a sociolinguistic investigation and aims to contribute equally to the fields of sociolinguistics, medical humanities, and cultural/media studies. The book is written in a context in which sociolinguistic work on the impact of pharmaceutical language is relatively scarce. Yet, several insightful scholarly contributions have been published since Tiefer (2006) observed that the non-pharmacological "study of drugs and society is still in its infancy but it will have to draw on multidisciplinary insights and skills from fields as diverse as science and technology studies, law, physiology, health policy studies, and the sociology of business and marketing" (p. 276).

In this medical humanities inquiry, conceptually, knowledge is presumed to "thrive(s) in the agency of culture and vice versa" (Oguamanam, 2008, p. 32). Throughout the book, the concept of knowledge is understood as a localized phenomenon, inherent to all societies, and tangible through material and immaterial expressions, awareness, and interdependencies. In the analysis that follows in the upcoming chapters, the sociolinguistic expression of the language of pharmacy is approached as the articulation, mobilization, documentation, and manifestation of a specific type of knowledge. Language signals pharmacological awareness (i.e., substance/precursor X treats symptoms/condition Y), and hence the uptake or archiving of information about medicinal substances, preparations, and practices through language. As excerpts from the corpora demonstrate, pharmacy-related by-products occupy a place in people's homes and conversations. Drugs are more than topics and references. They are epistemic objects with a place in the lives of their users (patients) and relatives. By-products of pharmacy are derivatives of science/knowledge, technology, and commerce, as well as sociocultural and linguistic intangible objects. As such, technology-derived objects are not just things (Friedel, 1993), certainly not when they are imbued with ideas that have epistemic implications (Herman, 1992). Materia medica are essential components of knowledge systems, shaped by local ontological perspectives. Including traditional and Indigenous knowledge within the notion of "local knowledge" is not a conceptual reductionism. Well-gatekept hierarchies of paradigms have perpetuated the absence of tools and insights from Indigenous epistemic systems (Chilisa, 2019), and this book does not intend to generalize or trivialize, the often cited (essentialist) binary distinction between Western and non-Western science.[57] Framed by the idea that pharmacy has accompanied humanity since ancient times, the central premise is that pharmacological knowledge has largely developed through trial and error, practiced in a localized manner according to specific times and places. The emphasis is placed on the linguistic dimension of pharmacological knowledge, as observed empirically in the consulted Icelandic and Spanish corpora. Selected excerpts are analyzed to infer how the language of pharmacy represents adaptable and localized knowledge about discoveries, novelties, and innovations.

As I will discuss in Chapter 4, the discovery of vitamins is one of those revolutions. Linguistically, the history of the synthetization, and classification of the 13 vitamins known to us today[58], showcases in interesting ways the relation between scientific language, commerce, and dissemination of information through media spaces. From a pragmatic perspective this is relevant in the context of register and the use of specialized lexicon. It is noticeable that in certain contexts, additional details about vitamers, or the generic descriptors are used instead of the shortened forms, for instance, the use of "riboflavin" and "ascorbic acid" instead of vitamin B_2 and vitamin C, respectively. More importantly, as a word adapted and adopted from science into everyday conversations gleans insightful aspects of the dynamics of linguistic elements derived from specialized terminology in non-scientific settings. An example thereof is "vitamin injection" (*Ic.* vítamínsprauta) to refer to an incentive or stimulus. I argue that the continuous optimization occurs *because* linguistic communities are not indifferent to the impact of science, technology, and epistemic additions and transfers. This is particularly noticeable in the case of the language of pharmacy *because* it is managed from its moment of conception (coinage) and until its moment of execution (adoption).

Similarly, I suggest that "containers", "labels" and "networks" can be helpful concepts to discuss the analogy I make between language and a logistic operation. Epistemically, I posit that "containers" encapsulate the volume of linguistic production that additions, shifts, or corrections in a knowledge system trigger and create. These would include the unstructured and structured input generated in a linguistic community when an epistemic alteration takes place (e.g., discussions about the possible existence of vitamins). Based on the examples from the corpora, in the case of Icelandic, the container for vitamins would include broad general definitions, such as "efni" (material/ stuff), or more specific proto-terms, such as "fjörvi", "vítamín- efni", "bætiefni", "fjörefni", "vítamina", "vitamin" and "vítamín". Nouns, either discarded, semantically redefined/reassigned or retained. While the official terminology for nutrition guidelines in Icelandic acknowledges both "vítamín" and "fjörefni", the MÍM 2022 corpus lists 14,557 entries for the former one, as opposed to 101 for the latter one. Following the logistic analogy, I suggest that epistemic labels emerge as the materialization of linguistic standardization (e.g., "vitamin" becomes a scientific term, *fjörefni/vítamín* in Icelandic, *vitamina* in Spanish). These are the denominations that language eventually adopts in varying ways and to varying extent. An example thereof is the less common use of "fjörefni" in the context of vitamins, probably due to its somehow broad meaning. Eventually, in logistic terms, I propose the term epistemic network to refer to the structures that allow for labels to become part of language. Through "networks" (e.g., a dictionary or media advertisement), labels become part of the linguistic repertoire of a community. Networks propel labels to reshape and expand the ways in which natural languages reflect the production of knowledge (e.g., adjective "vitaminized", *vítamínbættur* in Icelandic, *vitaminado* in Spanish). The central idea is that language can be viewed in logistical terms.

As for the selection of methodological approaches, my analysis is mainly informed by historical sociolinguistics, lexicography, language planning, loan word policy, the fringe areas of lexicography and folk taxonomy (Conklin, 1962, 1980), linguistic anthropology (Frake, 1977, 2012), cultural studies of technology, history of science, media studies, and development of terminology in the fields of chemistry, medicine, and pharmacy. The analysis in parts of the book is also illuminated by epistemic semiotics as a methodology. Packages, promotional material, and objects are used as outputs of knowledge production, diffusion, circulation, transfer, emulation, and manipulation. The above-mentioned notion of semiotic assemblages is also used as an entry point to the corpus of advertisements of medicines analyzed in Chapter 4. By this, I refer to Alastair Pennycook's approach in which "a semiotic assemblage expands the semiotic inventory and relocates repertoires in the dynamic relations among objects, places, and linguistic resources, an emergent property deriving from the interactions between people, artefacts, and space" (2021, p. 117). In this sense, "linguistic repertoires" are regarded as an "internalised individual competence or as the property of an imagined community" (p. 117).

A qualitative, and to a lesser degree, quantitative approach has been applied to the corpora, in addition to discourse analysis, morphosyntactic analysis, and corpus sociolinguistics. Besides historical sources, examples discussed in the subsequent chapters are extracted from a series of corpora of written and spoken Icelandic and Spanish: Corpus del Español NOW (Davies, 2016–2022) PRESEEA Corpus (Universidad de Alcalá, 2014–), Corpus del Diccionario histórico de la lengua española (Real Academia Española, 2013), Mörkuð íslensk málheild (Tagged Icelandic Corpus) (MÍM) (Helgadóttir *et al.*, 2012),[59] as well as the Ritmálssafn Orðabókar Háskólans database (Stofnun Árna Magnússonar í íslenskum fræðum, 2023). Reflecting on how written and oral corpora register linguistically the inception of medical, pharmaceutical, and ethnomedicinal epistemic additions, it becomes clear that language is not immune to the human dependence on knowledge about nature and its domains, the authoritativeness of science and the pervasiveness of culture. Besides textual and oral corpora (Icelandic and Spanish), linguistic data have been collected from written primary, audio-visual recorded material, digital media outlets, and digital linguistic databases and scientific linguistic planning documentation. As examples from natural language corpora indicate, these strikingly different approaches are reflected in the way in which speakers relate to the inception of scientific terminology. In combination with oral and textual corpora, other sources of spoken and written language include pharmacopoeias, pharmaceuticals catalog collections, Indigenous knowledge botanical compendia, academic journals, medical multi-terminology registers, textbooks, newspapers, magazines, televised material, oral history archives (i.e., ethnographical interviews), and radio broadcast recordings as sources of study cases.

Structure of the Book

As for the structure of the book, whereas Chapter 1 discusses the various ways in which the language of pharmacy can be conceptualized as a logistic task, Chapter 2, "Science, Materia Medica and Words" presents two study cases on the inception of chemistry, botany, and pharmacy in Iceland and Mexico. In the first part, "Chemical Sounds", I review the establishment of chemistry, as a scientific field and predecessor of pharmacy in Iceland in the 18th and 19th centuries and its direct relation to the dissemination of information about scientific discoveries taking place in main centers of knowledge throughout Europe. I trace the coinage of various terms suggested by prominent figures before "efnafræði" became the standard name of chemistry, showing the processes by which new vocabularies and terminologies are (un)co-ordinately managed to create linguistic conventions about scientific terms. In the second part, "Pharmacognosy and Local Knowledge", attention is drawn to the role of language management in the compilation of pharmacopoeias in Mexico between the 18th and early 20th centuries, specifically from the perspective of lexicon borrowed from Indigenous languages. As this chapter shows, language was crucial for Iceland and Mexico to mediate the incorporation, adaptation, and recognition of local pharmacy-related knowledges, but also the authority of science.

In Chapter 3, "Morphology and Pharmaceuticals", I review the construction of lexicon applied and applicable to pharmacy. In the first section, "Infixes, Identities and References", the main theme is the infixes **-ca-**, **-cur-** and **-quin-** as examples of traces of Indigenous local knowledges and languages in medicines developed mainly in the 19th and 20th century. Traces that are still prominent in Western pharmaceutical repertoires in substances found in most hospitals around the world. Focus is laid mainly upon the instrumentalization of language in the creation of a pharmacy-related vocabularies that reflect the categorization, properties, and origin of referents within predetermined taxonomies. As I show in the first part, "Infixes, Identities and References", names have an important history that situates and contextualizes forms and sources of knowledge. Language, as the examples illustrate, registers these processes. On this note, the next two parts, "Pharmaceutically Nominal and Nominally Pharmaceutical" and "Sociolinguistics of Pharmacy" advance further the discussion about the sociocultural and sociolinguistic implications of pharmaceutical names. Here, I discuss examples of innovations and discoveries (epistemic containers) that have materialized in terms and brand names (epistemic labels) intrinsically integrated into the language of pharmacy. To make this point, the chapter ends with examples of how names of medicines link past and present, local knowledge, and scientific endeavors, at times, in one word.

Bearing in mind the foundations of pharmacy as an amalgamation of science and commerce, Chapter 4, "Novelty, Lexicon and Medicines", revises the role of language historically in mobilizing pharmacy-related innovations. As mentioned before, in the first part, "In Terms of Science", I follow the emergence of the word "vitamín" and "fjörefni" in Icelandic, as two options for "vitamin" that emerged in the early 20th century in response to the discovery of "food factors". By reviewing the coinage of this term, I show how epistemic labels (e.g., vitamin) emerge from epistemic containers (e.g., understanding molecules in food). Particularly in the case of innovations that have revolutionized science in such profound ways, as the discovery of vitamins. In the second part, "Branding, Pharmaceuticalization and Translation", I explore the advertisement of pharmaceuticals from a linguistic and semiotic perspective. As the main study case, I analyze samples of medicines promotional material distributed among Latin American physicians in the 1950s, 1960s and 1970s. Dissecting the deployment of language in forms of medico-marketing allows us to make inferences about the role of commerce in the establishment and dissemination of pharmacy-related lexicon.

In the last chapter, Chapter 5, "Linguistic By-Products of Pharmacy", I study the adoption, appropriation, and use of pharmaceutical and medicine-related language in contemporary textual and oral natural language corpora. Here I show, that the commercialization of medicines has had a linguistic impact on Icelandic and Spanish speakers on various levels and in unsuspected ways. Whereas the first part, "The Sounds of Sexual Pharmacy" discusses the language of pharmacy in the context of reproductive and sexual health commodities, the second part, "Psychoactive Substances Slang", revisits the slang terms and cant used to refer to mind-altering substances in Iceland in the 1980s and briefly in the Spanish-speaking world in the 21st century. At last, in the third part, "Pharmaceutical Aural and Oral Side Effects", I discuss the impact of media in the promulgation of the language of pharmacy, showing that epistemic containers and labels rely on networks to be broadcast and communicated. In this final chapter, I demonstrate how the legal, economic, and biosocial aspects of medicines as substances for the care and cure of the body are adopted sociolinguistically in spoken language. As the excerpts of the corpora show, the linguistic implications transcend scientific considerations and regulatory protocols. The chapter conclusively reflects on how pharmacy-related jargon and branding propagate and prompt the circulation of pharmaceutical soundscapes across digital platforms in the 21st-century Iceland and the Spanish-speaking world. Conclusively, this book reviews how human communication moderates and reshapes the epistemic boundaries of new elements once they have been sociolinguistically appropriated and adopted.

Notes

1 Sukalo et al. (2021) and Cartwright and Armstrong (2020).
2 Blech (2023), Ledley *et al.* (2020).
3 This was already clear in the 19th century when Albert Lyons, the founding editor of *Pharmaceutical Era* (1887) noted in the first number of the trade journal: "The relations of the pharmacists to the manufacturer are becoming more intimate, and out of these relations there must arise many questions, economic and ethical" (p. 1).

4 Rightfully Haselow (2012) observes that: "Since the beginning of twentieth century the number of *nýyrði* ('new words') in Icelandic has been on the increase since one of the main challenges for the language planning institutions is to make the language cope with a globalized world (p. 146). As a language, it has a 'highly productive system of compounding'" (Whelpton, Guðmundsdóttir Beck and Jordan, 2015, p. 71), which has important implications at a morphological and semantic level.
5 Haselow (2012): "Apart from the language planning policies aimed at preserving a homogeneous standard language for all Icelanders a large part of the planning activities is devoted to the continuous modernization of the Icelandic lexicon" (145)
6 Hardon (2021), Van der Geest (2018), Romain (2013), Getz (2010), Giroux (2007), Stonier *et al.* (2007), King (2002), Whyte *et al.* (2002), Faure (1998), and Szasz (1976).
7 See for instance the dicussion in Hoffman (2017), Sierra Ospina and Tamayo Cano (2016), and Parkhurst (2007).
8 According to IFPMA, only in 2020, "53 new medicines were launched, while currently more than 9,000 compounds are at different stages of development globally", whereas in 2021, "the number of drugs in development for particular disease areas were: Cancer: 3,148, Immunology: 1,677, Neurology: 1,668 and Infectious diseases: 1,488" (p. 4).
9 Statista (2023) estimates that around 1.48 trillion U.S. dollars are spent in pharmaceuticals worldwide.
10 Language in biomedical settings might derail the understanding of patients about their health to the extent that the terminology can sound incomprehensible or even, foreign (Džuganová 2019; Peters, 2016; Bokhour *et al.,* 2015; Fields *et al.,* 2008). Jargon can also act as inhibitory factor in clinic settings (O'Connell, 2013).
11 See Kragh (2024).
12 O'Neil and Fetrow (2007).
13 In more specific terms, pharmacy in the context of Western medicine can be considered a design science according to the criteria stipulated by Niiniluoto (2014).
14 See the impact of Novo Nordisk's profits on Danish economy (Wass, S., Kresge, N. and Bloomberg, 2024).
15 As noted by Johnson (2016): "The pharmaceutical industry is trying, and succeeding, to work and profit in very different contexts, with very different regulatory frameworks, marketing needs and consumer bases" (p. 1).
16 See for instance Jakovljevic *et al.* (2021).
17 Human medicines are also some of the most strictly and highly regulated substances for human consumption on the planet. See Goedecke *et al.* (2018) and Matthews and Cartwright (1991).
18 See Gordin (2015, 2017).
19 See Sigurðsson (2010) on the role of Icelanders living in Copenhagen in the Icelandic Enlightenment.
20 See Luján-García (2020), Alcaraz Varó (2006).

21 From a history of science angle, Spanish is not categorically considered as a hegemonic language, despite its prevalence regionally, and political role within a colonizing empire (Gordin, 2017). On the language of science-empire dyad, see García Delgado, Alonso and Jiménez (2013), Cañizares-Esguerra (2006).
22 This could also reflect the continuity in the use of materia medica substances across the Mediterranean, the area having most impact on the Spanish-speaking world; see: De Vos (2010), Leonti *et al.* (2009) and Pollio *et al.* (2008).
23 In the sources, both "*Centraria*" and "*Cetraria*" are used.
24 See Martínez Solís, Aceves Pastrana and Morales Cosme (2007) on the impact of the pharmacopoeia in establishing pharmacy in Mexico by bridging policy, science, and education.
25 See Berg (1872).
26 Recent studies on its potential applications for the immune system have rekindled interest in *Cetraria islandica* (Iceland moss), as noted by Xu *et al.* (2016); Freysdóttir *et al.* (2008).
27 In *Diccionario de Ciencias Médicas* from 1824, a dictionary compiled by a "society of the most prestigious professors in Europe", the translators take the liberty of adding the following remark: "En España no se gasta otro liquen que el de la Península en razon de su abundancia y utilidad. Se encuentra en todos los puertos fríos, y particularmente en Asturias, Leyteriegos, y montañas de Leon: este es el mejor)" [*In Spain no other lichen is used than that of the Peninsula because of its abundance and usefulness. It is found in all the cold ports, and particularly in Asturias, Leyteriegos, and mountainous surrounding Leon: this is the best*] (p. 337).
28 See for instance the entry on Icelandic fjallagrös on the website of Slow Food Foundation for Biodiversity (2024).
29 Related to kava (*Piper methysticum*), which is currently experiencing a revitalization as a therapeutic used for relaxation purposes.
30 Etymologically, it derives most likely from Guyana Mukusi"wurari" (Burr and Leung, 2014). Variations in the name registered in European literature during the 19th and 20th centuries included "ourari", "urari", "wourara".
31 On the history of Introcostrin, see Ball and Westhorpe (2005) and Lee (2005).
32 Many are eponyms, such as biotite, boivinose, and friedelin (Senning, 2019).
33 See for instance the case of Korean revision and standardization of terminology, replacing the German-based system with an Anglo-Saxon one (Kim, 2022)
34 As well as the influence of Arabic on medical and pharmaceutical terminologies in other languages, see Kahl (2022), Al-Kasimi (2019), Darwish (2009), Huguet-Termes (2008), Norri (1991), Masliyah (1984), Levey (1962; 1973).
35 Akbari (2020).
36 Ivanova (2021).
37 See Cengizhan and Tanış (2010).
38 For the case of Benin, for instance, see Baxerres and Le Hesran (2011).
39 Sociolinguistic studies on the dynamics of Chinese, Japanese and Korean include Li (2022), Ahltorp *et al.* (2016), and Kim (2013),
40 See Singh (2022).

41 Some of the studied cases so far include Armenia (Portugal and Nonnenmacher, 2024), Azerbaijan (Valiyeva, 2021), Georgia (Rayfield, 2023), and Latvia (Apinis, Salaks, and Poznaka, 2010).

42 Notice the difference that Wenceslao Gonzalez (2021) makes between the language used for explanations and predictions (basic research) and that of predicting and prescribing, as is the case of pharmacology. Thus, "(f)rom this point of view, pharmacology does not investigate in the same way as genetics or molecular biology, although it needs the knowledge of genetics or molecular biology to make some its predictions about the ability to cure or mitigate a disease that can have some active ingredients of a medicine" (p. 15).

43 The history of medicine provides plenty of examples of references that have significantly evolved and shifted throughout time. As Martin and Martin-Granel (2006) have shown in their detailed study tracing the history of the term "epidemic", the historical implications on the semantic evolution of a term are major, and in the process, it is expected that form and meaning change, at least partially. See Mao and Fung (2020) on semantic referencing.

44 Other names of this opioid-receptor first synthesized in 1969 by Paul Janssen (Janssen Pharmaceuticals) in its generic formats include Diarfin, Elissan, Fortasec, Loperan, Loperkey, Protector, and Salvacolina.

45 Entry PR_32530395 (Corpus del Español NOW Corpus, 2016), originally collected from a Puerto Rican newspaper dated from 19 January 2006. The original spelling has been consistently preserved.

46 On the history and development of loperamide, see Van Rompay and Carter (1990).

47 This entry identified in the corpus was originally posted in an open discussion forum accessible to all users and unrestricted in terms of confidentiality and anonymity of the author of the post.

48 See Millones Espinosa and Hernández Maluenda (2020), Ortiz Gómez and Ignaciuk (2018; 2019), Gómez OrtPapí-Gálvez *et al.* (2007) and Sufrin and Ross (2008) on the impact of pharmaceutical marketing on women, Katsanis *et al.* (2012) on the complexities of determining the centrality of patients in relation to the commercialization of medicines, Sestino and Amatulli (2022) on the role of brand names in the choice of patients, Aufegger *et al.* (2012) on the relation between purchase intentionality and branding and Applequist and Gerard Ball (2018) on direct-to-consumer broadcast advertisement. For a historical overview of and study cases of pharmaceutical marketing strategies, see Buisan (2019), García Cárdenas (2019), Montero and Rodríguez-Martín (2019), Smith (2014), Núñez (2012), Kessel (2013), Fernández-Poyatos (2011). López-Muñoz and Alamo (2009), Watkins (2009), Crilly (2007), Chaganti (2006), Donohue (2006), McTavish (1999), Agrawal (1993) and Davenport-Hines *et al.* (1992).

49 On the relation between linguistic strategies and conveying the notion of medicinal efficacy, see Santamaría-Pérez and Marimón-Llorca (2008).

50 Sulfonamide medicines were first commercialized in 1930s as bacteriostatic and antimicrobials. See Bickel (1988). As a term in Spanish, "diyodohidroxiquinoleina" is first mentioned in the scientific literature in

Boletín de la Oficina Sanitaria Panamericana (1949) and a series of articles published in *Revista cubana de medicina tropical* (e.g., Castellanos, 1959).

51 By 1977, Vioftalyl was sold in jars and boxes of 24 tablets and the maximum price stipulated by the government would be 17,20 MXP and 10,90 MXP respectively (*Diario Oficial de la Federación*, 1977).

52 André Bigaux eventually became part of Sanofi in the 1990s (Gómez, 2002).

53 See the report prepared by Joan Brodovsky (1995) for the Economic Commission for Latin America and the Caribbean.

54 A medicine advertised as a panacea for exhausted men, based on an American formula containing iron and liver-based nutrients, widely advertised in the 1970s. See for instance the advertisement posted in *El Informador* on 6 August 1971 (p. 5A).

55 This is also the case in actual clinical settings unrelated to exposure to advertisement. LeBlanc *et al.* (2014) have shown the problematic aspect of overstimating "patients' understanding of medical jargon" (p. 241).

56 Exposure of readers to advertisement of medicines has nevertheless implications. An example thereof is Bigaux's advertisement of other of its products, Disprina, targeted at physicians, but published in a newspaper accessible to everyone: "Al H. cuerpo medico, André Bigaux, s.a. tiene el gusto de informarle que sus productos Disprina y Disprina Junior se encuentran a su disposición en todas las farmacias del país. Atentamente" [*To the honorable physicians, André Bigaux Ltd. is pleased to inform you that its Disprina and Disprina Junior products are available in all pharmacies in the country. Sincerely*] (*El Informador*, 16 August 1984, p. 6C). A product deliberately presented in a format suitable for children.

57 See David (2017) on the important aspect of objectivity in a discussion on Indigenous knowledge vis-a-vis science. See Gonzalez Rodriguez (2022a, 2022b) on the cultural perception of indigenous knowledge.

58 In their chronological order, the discovery of vitamin B1 (thiamine) (1910), was followed by the discovery of vitamin A (retinol) was discovered in 1912, vitamin C (ascorbic acid), vitamin D (calciferol) and vitamin B2 (riboflavin) in 1920, vitamin E (tocopherol) in 1922, vitamin K1 (phylloquinone) in 1929, vitamin B5 (pantothenic acid) in 1931, vitamin B6 (pyridoxine) in 1934, vitamin B7 (biotin) and vitamin B3 (niacin) in 1936, vitamin B9 (folic acid) in 1941 and vitamin B12 (1948).

59 Marked in this context follows the description of corpus in which "Með hugtakinu markaður (e. tagged) í lið (1)2 er átt við það að hvert einstakt lesmálsorð hafi verið greint m.t.t. tiltekinna málfræðiatriða oftast orðflokks og beygingarmyndar, og tengt við ákveðið uppflettiorð (lemmu eða les)" [*The term market (e.g. tagged) in point (1)2 means that each individual reading word has been analyzed with respect to of specific grammatical items, usually word class and inflectional form, and connected to a specific lookup word (lemma or lexeme)*] (Svavarsdóttir, 2007, p. 27).

1 Linguistic Logistics

The underlying premise throughout this book is that the language of pharmacy is logistically implemented and planned, but socially adopted, adapted, and appropriated. Both language and medicines move across societies in line with the precepts, conditions, and knowledge production and innovation of each time. I suggest that knowledge innovations (e.g., discovery of vitamins) act as containers from which new words, labels (e.g., vitamin C > ascorbic acid), enter networks (e.g., vitamin supplement add in the radio) that mobilize scientific concepts and terms that, once appropriated by speech communities, moderate and reshape the epistemic boundaries of new elements through daily communication (e.g., advertisement of *vitaminized* candy).

Humanity and pharmacy (not pharmaceuticals) share a long history. Pharmacological knowledge is embedded in most local medicinal knowledges in one way or another. How could we avoid talking about the things that we use to cure ourselves? Especially in cases in which it becomes a matter of life and death. Given that pharmacy is a science and body of knowledge focused on the preparation and administration of substances that have biological effects on living organisms, it clearly requires a technical language. Discussing medicinal preparations requires specialized concepts and terms. I suggest that language of pharmacy can be approached as a management process embedded in the organization and implementation of a complex operation: Communicating and registering the knowledge system of its speakers. From this perspective, language becomes a logistic undertaking. Like a parcel finding its way to its destination, linguistic units travel across a flow managed by the points of origin and ideally, points of reception, supported by the plasticity of natural languages to adapt and optimize themselves. As I show in this book, the language of pharmacy is planned and managed in similar ways as a logistic operation.

Even though all speakers of a language are not by default speakers of the various scientific versions of that language, in the 21st century, exposure

DOI: 10.4324/9781003601654-2

to terms and concepts derived from technological and scientific novelties is commonplace.[1] Although we might not all be fluent or eloquent, at least at a basic level, in the language of science, its presence in daily life is widespread. Chemical compounds listed in labels on products, instructional leaflets of devices and gadgets, information sheets attached to pharmaceuticals and revolutionary biomedical innovations or discoveries mentioned in the news are just some examples thereof. Not understanding each technicism does not prevent us from consuming products, using electric devices and/or if needed, taking medication. Quite the opposite, in paradoxical ways, our trust in science might have a lot to do with not understanding some of the technical terms mentioned in the media, product leaflets, and product labels. The linguistic distance between consumers and producers is symbolically indicative of a gap between levels of expertise linked to those derivatives of science and technology on which we appear to be dependent, for instance, medicines.

For most people, this has meant a major exposure to terms they might not be familiar with and many of those terms have *somehow* become part of their cultural repertoire. Since in present-day Western societies, health has come to imply more than just the state of not being ill, biomedical terms are nowadays, for instance, easily found on labels of food items, cosmetics, and toiletries. Language seems to moderate how medicinal efficacy is understood, perceived, and marketed. The ubiquitousness of health makes it attractive for its commercialization in all possible ways and that has implications for the general population.[2] Culture is not immune to the effect of science and scientific language is not immune to a process of cultural adoption and adaptation. Like any other logistic operation, biomedical language requires a stage of procuring, maintaining, and transporting, before it arrives at many of its possible final destinations (i.e., to serve its purpose).

1.1 Polyfunctionality of Language

Language serves many functions and has been studied from a myriad of points of view across time, space, schools of thought and through a broad range of interdisciplinary and inter-methodological inquiries and pharmacy is not an exception. In general terms, language is the main, or one of the main objects and subjects of study in linguistics and philosophy of language, linguistic anthropology, sociolinguistics, and communication. Language has been described as a system (Winograd, 1972; Savage-Rumbaugh, 1990; Zadeh, 2004), a representation (Gans, 1981; Blair, 1990; Grodzinsky *et al.*, 2000), a tool (Tylén et al., 2010; Everett, 2012), an instinctive manifestation of reality (Wittgenstein, 1953), an evolutionary ability (Gibson, 1991), a cognitive technology (Dascal, 2002), an entry point into the human mind (Dabrowska, 2004), an untestable ground (Troike, 1983), a cultural trait (Boaz, 1942; Jiang, 2000; Kramsch, 2014), a cultural barrier (McCarthy *et al.*, 2013), a social marker (Ireland and Mehl, 2014), an intercultural tool (Janík, 2016), a reflection of historical, social, political changes (Hock and Joseph, 2019) and a complex adaptation for communicative purposes (Pinker and Jackendoff, 2005). More pertinent in the case of this book, language has been described as an outlet for scientific expertise hierarchization (Levi Strauss, 1966; Hudson, 1978), a reflection of folk taxonomies (Conklin, 1962, 1980), a track record of the progress of human knowledge (Foucault, 1966) and a functional, nevertheless arbitrary classification instrument within medicine-related contexts (Bowker and Star, 2000). From within linguistic anthropology, recent scholarly work on pharmaceuticals has included studies on ethics and activism (Black, 2023), medicines and metaphors (Schirripa, 2015), local knowledge and meaning (Badie, 2013), biocommunicability (Briggs, 2011) and the veracity of claims (Wolfgram, 2010).

Yet, so far, limited attention has been devoted to the role of biomedical epistemic and technoscientific advances in shaping the capacity of natural languages to react, adopt, adapt to, accommodate, classify, and register the continuous evolution, production, and influx of knowledge on the speakers' own *terms*. Despite the crucial role of language in the documentation, categorization, and dissemination of scientific innovations, language has not yet been studied simultaneously from a logistic and reception-adoption perspective in the context of these novelties. On one hand, from a logistic standpoint, there is a lacuna in the literature about the ways in which the language of pharmacy emerges from logistic operations involving planning and execution in continuous optimization prompted by changes in the knowledge systems of its producers (e.g., scientific communities, pharmaceutical companies, regulatory boards), along with the history of these processes. On the other hand, from a reception-adoption viewpoint, scant attention has been devoted to the implications of biomedical- pharmacological language at a lexical, semantic, pragmatical and phonetic (i.e., phono-semantic matching) level as it progressively diffuses among the members of the recipient linguistic community in question. And which includes examples in the history of language and pharmacy about terms and names of medicines, preparations, and brands that were once used, but eventually discarded. Words that, perhaps cryptically, nevertheless offer important entry points into the language of pharmacy historically and the history of medicines from a (socio)linguistic vantage point.

While language has been recognized as a fundamental aspect of the human mind across various scientific disciplines, the impact of medical knowledge on language—and vice versa—has rarely been a central focus in extensive studies. The upcoming chapters aim to bridge this gap by exploring the language related to pharmacology and pharmacognosy as part of interconnected logistic (input) and reception-adoption (output) processes. By examining Icelandic and Spanish examples, I will demonstrate how technical language reflects the influence of science on its speakers and vice versa. Through an analysis of concepts, trademarks, and scientific terms (e.g., vitamins, Levitra, dope), the upcoming chapters will review how human communication moderates and reshapes the epistemic boundaries of new elements once they have been socio-linguistically appropriated. The main thesis posits that language transcends the linguistic boundaries established by science regarding the artifacts that embody the knowledge behind their creation.

With my discussion, I seek to add to the debate on the relationship between history of science,[3] cultural studies of science, scientific terminology and (socio)linguistics (Gordin, 2015, 2017; Bellés Fortuño, 2016; Degaetano-Ortlieb and Teich, 2022; Hepler and McEwen, 2019; Surman, 2019; Carli and Calaresu, 2007; Lonati, 2022). My goal is to offer a theoretical and yet empirically robust analysis built on excerpts of corpora from the perspective of historical evolution of language and historical sociolinguistics, in combination with contemporary natural language semantic, morphological, and pragmatic analyses. I depart from the notion that linguistic communities are also communities of patients and users of medicines for whom it is relevant to be fluent, at least partially, with the language of pharmacy. It can be a matter of life and death to know, for instance, the name of the diabetes-management medicine one takes every day. The everydayness of nouns, verbs, adjectives, and adverbs derived from science are focal points in my discussion. All along, observing though that these terms have been objects of rigorous managerial strategies. Scientific language is planned. Names of medicines are planned. The logistics of medicines is also planned.[4] None of this is new and none of it seems to change in the coming years or decades. Quite the opposite, the presence, pervasiveness, and prevalence of scientific language in non-scientific contexts is continuously increasing. The polyvalence of language allows it to embed scientific terminology (e.g., neologisms), adapting it and filtering it through idioms, colloquialisms, and jargon. The language of pharmacy illustrates the various fronts of the (socio)linguistic life of medicines and the coercion exerted on it from scientific and non-scientific corners.

From the 17th century, as noted by Halliday and Butt (2019) "vernacular languages in Europe, under this pressure from new forms of structured inquiry and debate, showed the effects of a new semantic pressure – the exposition of a logical (i.e., causal) sequence within a rhetorical posture of depersonalized objectivity" (p. 622). Besides the pressure to articulate, present and disseminate knowledge derived from scientific inquiry, European colonization also put pressure on language to mirror the material and epistemic exchanges, transfers, and additions that were taking place. The understanding of the world through natural sciences resided in inquisitiveness, as much as acquisition of goods. Objects, practices, concepts, and ideas from the Middle East, Far East, and the Americas also had a physical, cultural, political, and linguistic presence in the social landscape.[5] Additions in the form of *materia medica* brought from distant geographies had an immediate impact on the incipient forms of European pharmacy. And every "curiosity" had a name.[6]

André Thevet might have introduced tobacco into Europe,[7] but Jean Nicot's role in turning the plant into a medicine at the Medici's courthouse in the 16th century and Carl Linnaeus' binomial *Nicotiana tabacum* might explain why millions around the world are addicted to "nicotine", as opposed to "thevetine".[8] Neither is the substance name in any way related to "petun", "pitwna" or "pitima", some of the names used to refer to the plant in Brazil by the time Thevet *explored* the region between 1555 and 1556.[9] The Tupi-Guarani traces are not found in modern-day packages of antitobacco-addiction over-the-counter products. The language of pharmacy erased "petun". What stays is the confirmation that a series of logistic operations do determine how science and knowledge solidify linguistically.

As the explorers, religious networks (i.e., Jesuits in the case of cinchona) and courthouses from the past, today's scholars, and scientists, as much as influential figures within commercial networks of production and distribution have a major impact on the language we use in pharmacy-related contexts. Scientific publications have retained their status as the most salient source for the divulgation of terminology. While they also serve as platforms for negotiating the linguistic adaptation of scientific innovations. Scientists need language to communicate their findings, obviously. It is a matter of specificity, coherence, preciseness, and agreeability. In the case of pharmacy, trade names amount to a large proportion of the words identified in the various corpora consulted for this study. Due to their commercial nature, brands tend to be overlooked as samples of natural language. Yet, when a speaker describes an inspirational novel as "viagra de la literatura" (*Sp.* "literary viagra"), the reference only works at a pragmatic and semantic level because of the place of Viagra in the household medicinal repertoire. The meaning transcends the discursive boundaries of *the blue pill* as an erectile dysfunction medicine. Many other words can be used to imply that a work of fiction figuratively invigorates, energizes, or perhaps strengthens a withered literary scene. But Viagra is chosen for a reason, showing in the process the breadth and depth of pharmaceutical parlance beyond health-related contexts.

In parallel to semantic and pragmatic considerations, the implications of pharmacy in language can also be morphological and phonetic. In noting the eagerness of pharmaceutical manufacturers to create more memorable, distinctive, and intriguing names, Howard Ansel (1965) observes that, besides pronunciation, the secrecy behind certain brand names risks confusing physicians, druggists, and patients. Even in the pursuit of uniqueness, complex names force practitioners and pharmacists to risk becoming "amateur etymologists" (p. 160) in their attempts to familiarize or understand better the pharmaceuticals they prescribe and sell.

Understanding and familiarity are two key concepts to examine the role of language in pharmacy-related settings. On one hand, because the sites of scientific dissemination, divulgation, and sharing of findings, such as journals, conferences, or reports, revolve around comprehending science from a knowledge perspective (epistemic). The language is scientific to allow for the progress of science. Producers and brokers of knowledge (scientists, pharmaceutical companies, regulatory organizations, physicians, etc.) are bound to relate to language in sync with their scientific training or the professional settings in which they operate. On the other hand, patients' familiarity with a concept, a trade name, or a *long* name for a side effect does not imply they are conversant in the scientific language conventions of pharmacy or chemistry.[10] The polyvalence of language shows also through the various registers it allows according to the target audience of listeners and readers.[11] There are ways of *sounding* scientific without necessarily vocalizing, externalizing, articulating, or understanding the science behind a concept or a term. Halliday and Butt (2019) make a difference between "verbal science" and "verbal art". Whereas the second one deals with existential and moral issues through assessment and relationality, the first one invokes proportions and quantities "to explore the transactions between forms of matter and to capture the character and consequences of processes" (p. 624). The language of pharmacy, as far as terms and formulas is concerned, could be considered as "verbal science". The coinage of a trade name exemplifies the "verbal art" at best. Mainly in terms of its mere existence as a bridge that facilitates the application of biomedicine knowledge to improve human life. In logistic terms, the language conceived to describe, analyze, and disseminate pharmaceutical concepts and innovations is the starting point of a flow.

A supply chain of objects (medicines) and accompanying language that partially culminates in the reception of linguistic elements by the linguistic communities which include subcommunities of pharmaceutical users, promoters, brokers, sellers, etc. The linguistic-logistic supply line of pharmacy-derived objects and linguistic input never ends. Beyond the managed stages under control of pharmaceutical companies, doctors, and pharmacists, the biosocial presence of medicines in people's lives extends their leverage as part of therapeutic imaginaries. Oral and written natural language corpora attest and confirm in which ways pharmaceutical language becomes part of "verbal art" repertoires. In fact, the reaction of speech communities proves the logistic operations behind the mobilization of language is successful. As the last chapter shows, the language of pharmacy prompts linguistic manifestations in response to the ubiquitousness of medicines in modern-day life, or as I call them, pharmaceutical linguistic by-products. An example thereof would be "aspirin" in its diminutive form "aspirinita" (see example 1).

The logistics behind the overtly popular medicine has secured its availability, findability, and place in bathroom cabinets and street kiosks, not to a less degree than the leverage and recognition of the word aspirin itself. According to many dictionaries and spelling style guidelines, the word is not capitalized precisely because it is considered a simple noun. Few would ask for a tablet of acetylsalicylic acid. And yet, it is this term the one that renders aspirin understandable in the language of pharmacy. If indeed between 50 and 110 billion aspirin pills are consumed yearly around the world, aspirin must be one of the most recognizable words derived from science in current use.[12] Bayer has the credit for its introduction to the market in the 19th century, but also for coining a name vocalized thousands of thousands of times every minute with various local adaptations. **A-** (from acetyl), **-spir-** (from Spiraea), **-in** (suffix) is more than just a coined term. It is also the history of battles over patents, trademarks, commercial rights and of course, successful logistic operations to provide patients (consumers) with a pill to mitigate headaches and inflammation. As a word, "aspirin" exemplifies at best the linguistic logistics that applies to pharmacy-related nouns and terms. The history of the word itself preserves the marker of the language management behind its coinage. Ultimately, the first proposed named, "eusperin", was meant to reflect "Bayer's enthusiasm at that time for taking its tradenames from the Greek", but was eventually "rejected because it was felt that 'eu' was generally used to indicate an improvement in taste and smell" (Jack, 1997, p. 438). Eusperin never materialized linguistically. It was managerially prevented from entering our therapeutic vocabularies, for instance, in the treatment for "influenza" (Figure 1.1). Hence, the phonetics of marketing lies behind a brand name found in modern dictionaries across the planet.[13]

Since the first half of the 20th century, "aspirin" has been a product embedded in the logistics of circulation of medicines, but also, I claim, in the linguistic logistics of pharmacy and by proxy in the effect of science on language. How else would a word containing a reference to the Spirae genus be found in every corner around the world? And what does the infix **-spir-** tell us about the history of pharmacy and the linguistic marks left by pharmacognosy at the turn of the century? As noted by Halliday

Figure 1.1 Advertisement of Aspirin published in Spanish media in January 1918, the first year of the Great Influenza epidemic, also known as Spanish Flu epidemic.

Source: Photographer: Luis Bartolomé Marcos (2024).

and Butt (2019), science "has extended the resources of language by making novel demands on the meaning potential of the communities in which the practices of science have developed" (p. 622). Plenty of entries containing the word "aspirin" in the Icelandic and Spanish corpora confirm the universality of the antipyretic, analgesic and anti-inflammatory, commonly over-the-counter, medicine. As this aspirin poster from 1941 notes (Figure 1.2), when there is a risk of contracting a cold, aspirin is the "world famous remedy".

It is a matter of logistics that it can be purchased in airports, vending machines, and street markets, as well as a linguistic-logistic operation that "aspirin" (e.g., Korean 아스피린 a-su-p'i-rin), resonates with buyers in different languages. Before a process of labeling, the science behind the development of pharmaceuticals needs to be founded in the pharmacology of the compounds and molecules being used. Conceptually, as is the case of aspirin, a chemical formula [$C_9H_8O_4$] obtained from various stages of

Figure 1.2 Promotional poster of aspirin in Spanish. 1941.

Source: Photographer: Luis Bartolomé Marcos (2024).

experimentation and observation precedes the product that finally reaches the patients. In similar ways, I suggest that linguistic data travel across networks in the form of containers of concepts that are eventually labeled and categorized. Similar and in close connection with the language of chemistry, the language of pharmacy is managed, planned, and disseminated according to structured conventions. Experiments with substances leading to serendipitous results, such as a Viagra, can be described as accidental and anecdotal. But the language used in pharmacy-related contexts itself is not; not at all. Pharmaceutical companies design brand names and in line with regulatory bodies, follow instructions about the type of information that they need to provide physicians and patients. Nothing in such a careful line of production to distribution is a coincidence or an oversight and that applies to language as well. It is not a coincidence that "aspirin" would sound familiar to so many. It is a word derived from science, bearing the earmarks of the commercialization of pharmacy, but also of the operationalization of medicines as products that need to be understood or recognized by patients with no previous knowledge of either chemistry or pharmacy. Language adapts itself.

Aspirin becomes part of the therapeutic linguistic repertoire of millions and in the process an example of the logistic nature of pharmaceutical language. Not different from biomedical terminological approaches, language is managed to contain, contextualize, and categorize information in a precise, efficient way. Yet, chemical, and pharmaceutical languages differ saliently in two aspects. On one hand, trade names (trademarks, brand names, etc.) are specifically coined for commercial purposes. On the other hand, the language used to promote innovations in pharmacy is nevertheless parsed in terms of chemical terminology. But another aspect is even more important for any discussion about language in pharmacy-related settings. That is the biocultural and sociolinguistic adaptation of medicines as part of daily conversations, informal speech, street language and even slang, as I will eventually show in the subsequent chapters. Once the language of pharmacy has found a place in the vocal cords of the patients, medicines and biomedicine-derived vocabulary become part of natural language. The full version of the previously mentioned example (i.e., aspirinita) from the NOW Corpus in Spanish illustrates my point:

1 El deporte entra ahí como una aspirinita de salvación (Mexico 2018).
[*Sport becomes part of it just as a life-saving little aspirin*]

The use of diminutive (**-ita**) in Spanish is observed to convey a sense of familiarity, appreciation, connection, affection and at times, paternalization (i.e., the language used to address children).[14] A diminutive affix also can articulate disdain, apprehension, or lack of appreciation, or at the same time, mitigate them (García, 2016). As a descriptor, in less semantic-pragmatic terms, it implies the noun in question refers to a small person or object, or at least, smaller in comparison to the interlocutor.[15] Both quantitatively and qualitatively, **-ita** in this context, indicates a certain degree of closeness, and fondness of a small amount of a substance (aspirin) from which the interlocutor benefits figuratively. "Little aspirin" denotes for the speaker a pill of smaller size than usual that does wonders, probably in similar ways that an aspirin, the pharmaceutical type, alleviates headaches and relieves inflammation and body aches. It is a brand name, a generic medicine[16] and a substantive, but also a by-word that semantically conveys the knowledge and possibly lived experience of the speaker who utters it. At the reception end of the supply chain of medicines, pharmaceuticals become more than lemes or rhemes, they are topics of conversation (in their own right).

Aspirin and many other similar examples illustrate the response and adaptation to the language of pharmacy in everyday speech, but also their adoption along with references that are meant to *sound* scientific. Yet, many of which obscure the traditional knowledge behind the creation or epistemic origin of some of the substances we colloquially called "drugs". Returning to the history of aspirin, for example, is after all returning to the history of salicylic acid, obtained mainly from myrtle and willow bark leaves, and its use for thousands of years in therapeutic preparations as part of Ancient Chinese, Assyrian, Greek, Native American, Roman and, Sumerian civilizations.[17] To name one of the oldest sources, willow is mentioned as an analgesic and anti-inflammatory in Ebers papyrus, an Egyptian herbal knowledge dating to ca. 1550 BCE. The chronicle of salicylates and acetylsalicylic acid, its derivate, includes recommendations made by Hippocrates about the use of willow tree (*Salix alba*), reiterations by Plinius and Dioscarides about its medical properties, and thousands of scientific articles describing episodes of innovation, trial and error, and constant furtherance and experimentation. More relevantly for the linguistic discussion is the fact that aspirin as a "household remedy" (Schrör, 2022, p. 22) quickly became an integral part of pain and inflammation management in the early 20th century. It had rapidly "become a drug whose name was not only well known to health professionals but also to the general public – and accepted without reservation by both" (p. 21). Aspirin was an epistemic and material addition at a biosocial level, as an analgesic, but also a linguistic addition as a byword for a technoscientific therapeutic novelty.

Aspirin and many more examples examined in the next chapters illustrate how linguistic elements conceptually derived from biomedical settings have the capacity of being naturalized and incorporated in natural language in unexpected ways. Thus, elements from pharmaceutical speech are linguistically accultured; after all, they serve communities of living speakers in constant change. Quoting the wisely put observation by Enfield and Sidnell (2022): "Language doesn't just allow us to navigate a social environment, it transforms and indeed constitutes the very social environments we must navigate. This is one of language's most momentous consequences" (p. 186). I celebrate the use of "navigation" in the context of language as it is useful to understand my logistic-linguistic analogy. As much as words move across oceans of vocabularies and terminologies, they create waves that eventually materialize in additions and changes in the language of their speakers. Whereas elements in natural languages cruise, I would argue—metaphorically—terms derived from scientific language schematically sail, at least at first, across the open seas of human communication, only to adapt and eventually, perhaps, change course.

Scientific language is logistically planned, yet socio- linguistically appropriated and adapted, nonetheless in the case of words that articulate matters of life and death and health and illness, such as pharmaceuticals and biomedical concepts. From this angle, one can say that language embeds science, while scientific conventions structure language. A language that when and if needed, societies appropriate and adapt according to their needs. When a speaker refers to doing sports, for instance, as an "aspirinita" (*Sp.* little aspirin) for one's wellbeing, the diminutive form of aspirin does not hint at the efficacy of acetylsalicylic acid. Neither does this reference consider the nonmedicinal ingredients of an average aspirin (i.e., corn- starch, Hypromellose, powdered cellulose and triacetin). For this speaker, "little aspirin" denotes efficacy and familiarity. After more than hundred years, aspirin has ceased to be a coined brand name or a trademark.[18] It is a reference, a referent and a science-derived noun engrained in the linguistic repertoire of millions, after a process of coinage, planning, management, and dissemination. As many other words, "aspirin" was scientifically, linguistically, and logistically devised and as I show in this book through various examples, natural language has not remained immune to the inclusion of science-derived words.

Historical sources have registered the various ways in which language has played a crucial role in facilitating the exchange and mobilization of scientific information and its presence in public discourse, particularly in the 20th and 21st century.[19] But how did scientific language emerge? And what does this mean for language produced in scientific settings once it finds its way into non- scientific contexts? And who and what controls how science is expressed linguistically and where it is used? There are not always straightforward answers to these questions. On the one hand, because science depends on the capacity of language to allow for it to be communicated ideally with precision and clarity. It is in the scientists' best interest to maintain their specific use of language at all cost. On the other hand, because those fluent in a particular terminology are trained to be so and there is no need for the entire population to be familiar with every technicism that exists (e.g., technical names of diseases and symptoms). It would be an impossible task.[20] Scientists themselves are for the most part only familiar with the jargon specific to their own field. They use and produce their own version of language, understanding it as (one of) the main methods of communication between humans, structured in such a way that it conveys meaning through corporal, oral or written forms. The last two constitute the raw material for the argument advanced in this book.

As for pharmaceuticals, they are technological outputs[21]. They might be mass produced, but at the outset, humans design them for other humans to use them to treat disease. Even with the ever-growing intervention of artificial intelligence, the knowledge production behind the pharmaceutical industry, human actors will probably continue intervening in the design of medicines. At all stages, the adaptability of language makes these cycles possible. The inescapability of communicational tokens, signs, or characters, even at the simplest level, prompts science to maneuver language in sync with time and space. Not the other way around. Language serves science. But language serves also culture through the interconnected social fabric of large (e.g., Spanish) and small linguistic communities (e.g., Icelandic). They all rely on a type of pharmacy for their health and on language to communicate about it.

1.2 Logistics and Linguistics

By allowing us to communicate with each other, language efficiently fulfills its main function. It moderates the complex operation of transporting thoughts from one mind to that of others by means of sound or graphic symbols. Linguistic structures demand complex computations. The words need to be in such an order that they convey the intended message and eventually produce the desired outcome. They also need to be specific enough to avoid ambiguity, although that is not always possible, but at least to the degree needed to allow for mutual understanding. The detailed planning behind the implementation of language as an operation, I suggest, resembles the management needed to complete a logistic task. Logistic as the art or procuring, maintain and transporting A from B to C is not very different from language conceived in simple terms as having B saying A to C.[22] In general terms, logistics has so far been conceptualized for the applicability of models to undertake operations (Klaus, 2009), updatable with the advent of newer technologies (Ding *et al.*, 2021). From a linguistic perspective, the intersection between language and logistics has mainly been revisited in terms of action-based processes and implementation of models, classifications, and ontologies (Küng, 2012; Thiers, 2011). From a logistic-only perspective, language is regarded as a structuring framework.

Yet, language itself serves as a reservoir of human knowledge. It is a fundamental tool in communicating information about the natural world and preserving accumulated knowledge from generation to generation, nonetheless on how to enhance, improve and restore human health. Language has been useful in reporting on informal observations, but also in registering the findings derived from formal trial-and-error experimentation. Either in an unstructured or structured fashion, there is a strong link between language and the logistic dissemination of knowledge from point A to point B. It probably stems from the fact that sharing information requires finding an efficient route (e.g., print media) and a common ground to do so, especially when it is a matter of survival, such as the case of health-related knowledge. In the 21st century, talking about health often involves the inclusion of biomedical terminology, even if only partially. This is not new. Long before the onset of chemistry and pharmacology, everyday parlance about remedies, cures, and concoctions required the use of specific words. In modern times, a conversation about illness and medical treatments would certainly include terms related to therapeutic interventions and certainly pharmaceuticals. Both language and logistic operations take place in enclosed systems interconnected by systems that facilitate the flow of living organisms or supplies in the first case, words in the second case.

For the specific situation of the language of pharmacy, one thing is clear, many of the words derived from scientific terminologies are conceived, coined, and managed long before they become part of natural languages. Ibuprofen does not have an etymological past that can be contextualized in terms of linguistic relatedness, century-old important cultural exchanges, religious, or political common history, or an anecdote. It is a brand name derived from the terms isobutyl, propionic acid, and phenyl, dating from 1960s in a British laboratory. Historically, ibuprofen has a stronger nexus to attempts to create a pill to mitigate hangover than to a crusade, a network of devoted scholars based in Medieval universities or any process of linguistic colonization—as is the case with many medical words of opaque precedence. Ibuprofen was logistically conceived to become part of the English language. In sync with the nature of pharmacy as a science-industry, both science, and commerce lie behind the coinage and circulation of $C_{13}H_{18}O_2$, which would be the formula chemists would use to identify its composition.[23] The nominal identity of ibuprofen (i.e., 布洛芬 Bù luò fēn in Chinese) gets more complex if one considers that it is sold under 60 different names worldwide.[24]

Returning to the examples of ibuprofen, as part of natural language, ibuprofen ceases to be a long four-syllable unpronounceable word with a strange beginning and ending.

2 Estoy mareado, me duele la cabeza… .¿tenéis un ibuprofeno?
– No, lo siento (Spain 2019)
[*I'm dizzy, my head hurts… .do you have an ibuprofen?*
– No, I'm sorry]

3 Los versos pueden ser auténticos ibuprofenos, alivio rápido y sintomático contra.
el dolor, la desesperanza, la tristeza (Spain 2019)
[*The verses can be authentic ibuprofen,*
fast and symptomatic relief against pain, hopelessness, sadness]

Evidently, 2 differs from 3 mainly at a pragmatic level with the first example alluding to the nonsteroidal anti-inflammatory drug, while the second to the symbolism of modern-day semi-panacea for spiritual conditions. In 2, the speaker is acquainted with the renown efficacy, pharmacodynamics, and pharmacokinetics of the medicine. The sense of time is implied in dizziness and headache that need immediate attention. Temporality is also implicit in 3, specifically emphasized by the "rapid relief" that the discussed verses can release against emotions, such as a hopelessness and sadness. Ibuprofen in its plural form (*Sp.* ibuprofenos) operate here as synonymous for "remedy", "antidote", "boost", "cure", "countermeasure", simply put, "solution". Or as in this entry from the Icelandic MÍM corpus:

4 Íbufen er besti vinur þinn akkurat núna og klaki (2010).
[*Ibuprofen is your best friend right now along with ice*]

Progressively, pharmacological knowledge is adapted and adopted in everyday life through language. Science-related elements become part of our vocabularies. Before "íbufen" (*Ic.*) and "ibuprofeno" (*Sp.*) became a *household* name though, it had to exist as a *name* at all.[25] It had to be invented, managed, and mobilized from the experimentation setting of a laboratory, via the catalogs of pharmacists and into familiar settings in which it could pragmatically be accommodated with connotations such as "antidote", or a reference related to friendship in any figurative way.

Regardless of the language in question, ibuprofen has become part of the linguistic pharmacological repertoire, for instance, of those wanting to relieve fever. "Íbúfen" (*Ic.*) and "ibuprofeno" (*Sp.*) refer to the same substance, and yet they move across very different and yet similar linguistic communities. Both Icelandic and Spanish have gone through various processes of adapting and accommodating linguistically scientific innovations developed in other countries. In both cases, translation of foreign medical literature played a crucial role in the establishment of local scientific communities.[26] Reactions to the inflow of loanwords and foreign neologisms were also similar. It prompted a sense of awareness about the need to create protocols for scientific terminology. Protocols that reveal that scientific and health- related language undergoes indeed processes of logistic management that contain linguistic elements derived from traditional knowledge/ heritage (i.e., quinine in the cabinets of 16th-century-apothecaries; quinoa (*Chenopodium quinoa*) in the shelves of 21st-century health stores and organic product corners in supermarkets).

If linguistics is the systematic analysis of all aspects of language in a comprehensive way, there could be room in the field to consider the ways in which certain types of language emerge from logistic processes. As I suggest in the subsequent chapters, the language of pharmacy illustrates at best the process of creating and upholding a linguistic system built on a flow of stages. All of which ultimately result in the completion of a task. The previously mentioned example of ibuprofen crystallizes the various steps needed before this term enters the therapeutic imaginary of a speaker suffering from headache. Knowledge production (science) preceded a discussion about the term (science and commerce), which subsequently preceded the circulation of the word (commerce and marketing) and its eventual establishment in modern-day English, before it would become part of hundreds of other languages across the globe. Icelandic and Spanish, as the entries from the corpora show, are two of them. Another commonality is that both in the case of Icelandic and Spanish, there seems to be the belief that there is such a thing as "Standard Spanish" and "Standard Icelandic" (Vogl, 2012, p. 10). As Vogl (2012) observes about Standard Icelandic, there is "a rather strong, shared belief that there is a one-to-one relationship between nation, speech community, and language – Icelandic as the language of the Icelanders in Iceland" (p. 10).

In the case of advances in pharmacology, Iceland, Latin America, and Spain, with some few exceptions, were more at the receiving rather than entrepreneurial side for most of the 19th and early 20th centuries. In terms of points of divergence, Icelandic, and Spanish have had very different language planning policies in their adoption of biomedical terminology.[27] Whereas Spanish has adopted many terms from French, especially in the 19th century, English, and German in the 20th century and English in the 21st century, purism has only partially been advocated for and uniformly implemented across the Spanish-speaking world. In Iceland, on one hand, "a core theme of Icelandic language ideologies is grammatical, orthographic and lexical purism" (Kristinsson, 2018, p. 246).[28] On the other hand, the creation of centers of formal education (Ottósson, 1990) and the involvement of prominent figures (See Chapter 2), along with the independence campaign against the Danish Crown had a long-standing impact on the country's language policy. Haselow (2012) notes that the "spread of potential sources of linguistic variation is regularly slowed down and, if possible, stopped whenever it affects structures on the phonological or grammatical level" (p. 147).[29] Both in the case of Icelandic and Spanish, geographical proximity to major industrial centers of knowledge production has had an impact on the incursion of scientific terminology.[30] Spain developed through a major part of the 19th and 20th century with France a similar relation Iceland had with Denmark. In both cases, the impact of German major advances in the first half of the 20th century prompted important lexicological exchanges (See Chapter 2), sometimes in unexpected ways.

In current times, similarly, the increased use of English as lingua franca in medicine, chemistry, and pharmacy has shaped unprecedentedly the influx of scientific terms derived from anglicisms, directly translated, or morphologically adopted.[31] Hoffman (2012) alerts us though about the importance of realizing that many scientists fall into the illusion of thinking "that what they say is not influenced by the language they use, meaning both the national language (German, French, Chinese) and the words within that language" (p. 153). This is a particularly relevant observation in the case of languages that have operated historically more at the reception rather than creation side of scientific innovation. Moreover, this is an important point from the perspective of terminology, but also of the registers used to inform users about the medicines they take, what they contain and how they have been developed. On this, Grabowski (2015) notices that "pharmaceutical discourse does not constitute a single, homogenous and uniform linguistic phenomenon" (p. 24), rather an agglutination of various discursive styles. A relevant aspect to consider, for instance, in the context of the marketing and advertisement of medicines across media outlets. About words, Hoffman (2012) warns us about the belief that "providing they're well-defined, [they] are just representations of an underlying material reality which they, the scientists, have discovered or mathematicized. Because the words are faithful representations of that reality, they should be perfectly translatable, into any language" (p. 153). Crosslinguistic translation in scientific domains, as it has been consistently shown, follows specific procedures. Many pitfalls and obstacles in transferring concepts across linguistic barriers indicate Hoffman is right about the real implications of language from within science. In sum, the managerial dimension of transposing terminologies from one language (often a dominant one) to a set of others is a logistic enterprise—in its own right. It is part of a flow, I suggest, of steps that secure the completion of the main task, which is that of constructing and maintaining a language about and of pharmacy and knowledge about medicinal preparations.

If indeed the language of pharmacy can be conceived as a logistic operation, I suggest comparing its various stages to the use of containers, labels, and networks. Specifically, throughout the book, I work with the notions of epistemic containers, epistemic labels, and epistemic networks. The key concepts in a linguistic logistic approach are epistemic containers, epistemic labels, and epistemic networks. By epistemic containers, I primarily refer to sites that encapsulate the organization of scientific language and reporting venues of traditional knowledge. These are constituted by the descriptions, discussions, controversies, interactions, and negotiations registered by language every time there is an alteration or addition in a knowledge system. Containers are reportorial and hold a historical value to understand how the language of pharmacy has emerged (e.g., scientific journals). When agreement has been reached, labels emerge. Epistemic labels are the actual designators and traces of knowledge institutionalized by language. These are the solidified references and linguistic standardized denominators used to create structures around epistemic changes and innovations. In the case of pharmacy, commerce plays an important role in the creation of labels through the coinage of brand names, devised specifically to draw attention and create the linguistic identity of pharmaceuticals (e.g., aspirin). Subsequently, labels enter organized networks. These are complex webs of information that secure the flow of knowledge between the point of origin and the recipient linguistic communities. As mentioned in the introduction, from this angle, while "vitamin" would be an epistemic label, "trophone", "sitacoid", "vitaaids", "biogens", "advitants", "ergotropes", "ergogents" and "accessory food factors", discussed at some point in history as possible scientific terms to refer to vitamins (Ramsden, 1918; Brown, 1920; Maltz, 1992), would be pieces of an epistemic container.[32] In this context, a modern English dictionary with the entry "vitamin" would be a good example of an epistemic network.

Once a term, such as "vitamin" is adopted by a community of speakers, the managerial aspect of the linguistic- logistic flow ends. Throughout the book, I use the term linguistic health-spheres to refer to that space in which the implications of the appropriation and adoption of pharmacy-related lexicon takes place. New structures and forms emerge, even if they might obscure the knowledge behind them, as it has been the case in English with, for instance, "vitamin", "vitaminic", "to vitaminize", "vitaminized", "multivitamin", "antivitamin", "megavitamin", "provitamin", "avitaminosis" and "avitaminotic". These are by-products of the linguistic-logistic process that the discovery of vitamins activated since the early 20th century (See Chapter 3). Within the category of linguistic by-products, I situate all the production of natural language that has emerged in response to the incursion of pharmacy-related vocabularies. Despite its relative triviality, I posit that the mere existence of the above-mentioned "aspirinita" (*Sp.* little aspirin) has many lessons to teach us about the interaction, impact, presence, and reception of science in language. To illustrate this point better, consider the following examples of the use of the verb "inyectar" (to inject) in Spanish found in the NOW Corpus:

5 Se enfoca en inyectarle a su personal las ideas de responsabilidad social y superación, tanto dentro (Colombia, 2014).
[*It focuses on injecting its staff with the ideas of social responsibility and improvement, both within*]

6 Está con nosotros y con lo que yo le pueda inyectar a los jugadores anímicamente va a ser bravo ganarnos (Argentina 2018).
[*He is with us and with what I can inject the players emotionally it will be brave to beat us*]

7 Luis Enrique quita trascendencia a el enfrentamiento para no inyectar más presión de la necesaria en el vestuario (Spain 2014).
[*Luis Enrique minimizes the importance of the confrontation so as not to inject more pressure than necessary in the dressing room*]

8 Para eso lo mejor es el láser o inyectarte ácido hialurónico. Puedes consultar a el doctor (Spain 2013).
[*For that, the best thing is the laser or inject yourself with hyaluronic acid. You can consult the doctor*]

9 En el funeral tuvieron que inyectar un sedante a la viuda, que no soportaba la tensión de el momento (Spain 2013).
[*At the funeral they had to inject the widow with a sedative, who could not bear the stress of the moment*]

10 La epidural suele inyectar se por medio de un catéter durante el parto, pero generalmente los médicos (Mexico 2017).
[*The epidural is usually injected through a catheter during childbirth, but usually by doctors*]

11 Los medicamentos de quimioterapia se pueden inyectar en una vena o se pueden tomar por vía oral (Peru 2019).
[*Chemotherapy drugs can be injected into a vein or taken by mouth*]

Reviewing these examples, it becomes clear that "inyectar" moves across different spaces with various connotations and registers, as it is the natural course of words from a pragmatic, and at times semantic perspective. But what 5, 6 and 7 do show is the scientification of colloquial language in informal settings, sports, and personal preferences in these specific cases. The terms "inyectar" (to inject), "inyección" (injection) and "jeringa" (syringe) have a relatively long history in Spanish. As an entry in the *Diccionario de Autoridades* (1734), "inyección" is described as a medical or chemical term, referring to the action of introducing or pouring a liquor onto another body by means of an instrument, derived from "Injectio" (L.) (Volume IV). In the VI volume, "xeringa" (syringe) is described an:

> Instrumento compuesto de un cilindro hueco, ò cañón gruesso de metal, en que está unido otro cañoncito mui delgado, por el qual se atrahe, y arroja el liquór, al modo de la bomba, ù otras máchinas hydráulicas, con la fuerza, y artificio del embolo, dirigiendole à la parte que se quiere, para los efectos, que se intentan; à proporcion de los quales es tambien el tamaño de dicho instrumento. Díxose del Griego Siringa, que significa caña, ò cañón; por lo qual debe escribirse con x. Lat. Sipho, nis. Siphunculus, i. Clyster, eris. [*instrument made up of a hollow cylinder, or a thick metal cannon, in which another very thin cannon is attached, by which the liquor is drawn in and expelled, in the similar to a pump, or other hydraulic machines, with force and artifice of the plunger, directing it to the part desired, for the effects intended; the proportion of which is also the size of said instrument. Derived*

from the Greek Siringa, which means reed, or *cannon; so it must be written with x.*]

(1739)

Based on this, the Greek spelling would have been from Greek ὑγριγξ (syrigx). In the *Nuevo Diccionario Latino-Español* (1867), compiled by Raimundo De Miguel and Joaquín Gómez de la Cortina (Marqués de Morante), "injectĭo" is defined as "La acción de agarrar, de tomar posesión" [*The action of seizing, of taking possession*] (p. 471), attributing Fabius Quintilianus as original source. A related entry, "injectus" is defined as the action of throwing onto (quoting Pliny) or else the action of covering (quoting Papinius Statius and Cornelius Tacitus) (p. 471).

Whereas it could be argued that the figurativeness of "injecting ideas", "injecting someone emotionally" and "injecting pressure" can easily be explained from a pragmatic angle, I argue that the notion of "injecting" retains its linguistic leverage mainly due to its sustained discursive presence in relation to medicines being injected. In the subsequent chapters, similar excerpts from the corpora are used to showcase how Icelandic and Spanish have diachronically (1850–2022) adopted and registered the incursion of pharmacy-related knowledge, technologies, and commerce.

Notes

1 See Thorakkattu (2022), Schwartz and Woloshin (2019), Shanahan (2019) and Breitenback and Fischer (2013).
2 See Fochler (2016).
3 Chemla and Virbel (2015).
4 Throughout the book, logistics is defined as the meticulous planning and execution of a complex operation. Over recent decades, this term has increasingly been adopted and used beyond its original military context, see Lynn (1993), Schouten *et al.* (2019), Cowen (2014), Bamyaci (2021).
5 See for instance Mikaili (2012) and Fernández Rodríguez (2023).
6 The term "derives from an influential text of the first century BC by the Greek physician Pedanius Dioscorides, *De materia medica*" and which "has come to designate knowledge of all kinds concerning the medicinal properties of natural substances, and their preparations and use in healing" (Makowska-Wąs *et al.*, 2022, p. 275). De Vos's (2010) quantitative study indicates that "while Hippocratic medicine may have provided an early basis for knowledge and use of materia medica, it was definitively superseded by the work of Dioscorides in the first century A.D" (p. 47), which probably contextualizes better the retention of the term.
7 Thevet could have been the first one to write about tobacco for European audiences, not necessarily the first in introducing the plant to the French.
8 Perhaps to Thevet "rightfully belongs the credit of having introduced tobacco into France. But NICOT'S later introduction of the same plant, his experiments with it in Lisbon, his letter to the Cardinal of Lorraine, and his sending the plant to CATHERINE DE' MEDICI were more spectacular than anything THEVET did to make himself known" (Bowen, 1938, p. 361)

9 See Laufer (1924).
10 On the history of linguistic conventions in chemistry and pharmacy in Icelandic and Spanish, see Gylfason and Gylfason (1985), Rodríguez González (1994), Brumme (2001), Bertomeu Sánchez and Muñoz Bello (2010; 2012), Pinilla and Lépinette (2015), Muñoz Armijo (2018), Álvarez Jurado (2021), Bautista Becerro (2022), Kristinsson (2014; 2018; 2022).
11 Pieterse *et al.* (2013) alerts us to the importance of taking into consideration the educational background as an important aspect to assess the implications of opaque terminology in medical settings, as well as competence in the language (Hayes *et al.* 2017). A factor that most likely also has an impact on how potential patients/users understand terms used in advertisements of medicines. On the readability and semiotic legibility of leaflets, see Cohen and Moliner (2022), Medina-Córdoba *et al.* (2012), and on leaflets and information about medicines on digital platforms, see Young *et al.* (2019).
12 See Marko (2020).
13 But commerce plays of course a minor role in other scientific fields when it comes to establishing a fixed terminology. Scientific terms evolve. An example thereof could be "korgvikindi" for insect, as opposed to "skordýr" in Modern Icelandic (as registered in *Ordabok, sem inniheldr flest fágiæt, framandi og vandskilinn ord, er verda fyrir i dønskum bókum* by Gunnlaugur Oddsson, 1819).
14 See King and Melzi (2004) for their analysis on the use of diminutives in mother-child interaction.
15 Considered the most obvious and idealized use (Malaver, 2018; Mendoza Ibáñez, 2000).
16 See Greene (2014).
17 As noted by Jack (1997): "Human beings have suffered from inflammatory rheumatic disease since ancient times. The Assyrians left behind stone tablets from the Sumerian period describing the use of willow leaves for this condition. The Egyptians were aware of the analgesic effects of decoction of myrtle or willow leaves for joint pain and the Ebers papyrus has a very accurate description of an inflammatory condition: "When you examine a man with an irregular wound ... and that wound is inflamed ... [there is] a concentration of heat; the lips of that wound are reddened and that man is hot in consequence ...then you must make cooling substances for him to draw the heat out ... leaves of the willow" (p. 437).
18 Schrör (2022) reminds that: "Aspirin is not only one of the most intensively studied but also one of the most frequently used drugs worldwide, today with an impressive annual production rate of 8 billion tablets alone in the German Bayer plant. About 2,500–3,000 entries for the term 'aspirin' (acetylsalicylic acid) can be found in the PubMed database – every year. All this happened more than 100 years after the first pharmacologist studying the compound supposed 'the substance is of no value' and the first clinician who used the drug for treatment of inflammatory pain did so with 'not little distress'. We all know now that the reality soon became another one" (p. III).
19 See Nieto-Galan (2016) on the historical interplay between media and science popularization.

20 What can represent a problem for patients, as LeBlanc *et al.* (2014) has shown in regards to the opacity of bio-medicinal jargon in clinical contexts.
21 See Mikaili (2012) on other ways of understanding pharmaceuticals.
22 See for instance the Oxford English Dictionary definition, according to which, logistics is "the branch of military science relating to procuring, maintaining and transporting material, personnel and facilities" (2022). This differs from, for instance the New Oxford American Dictionary entry for logistics, which defines it as "the detailed coordination of a complex operation involving many people, facilities, or supplies" (2022).
23 Hoffman (2012) alerts us nevertheless of the limits of language, even chemical one: "That position is defensible—as soon as the synthesis of the new high-temperature superconductor $YBa_2Cu_3O_{7-x}$, was described, it was reproduced, in a hundred laboratories around the globe. But the real situation is more complex. In another sense words are all we have. And the words we have, in any language, are ill-defined, ambiguous" (p. 153).
24 At an international level, supranational organizations are invested in creating and mobilizing nomenclature systems for the names of medicines. See for instance the full statement of the World Health Organization (2017) in regards to International Non-proprietary Names (Appendix 1).
25 See Murphy (2018), Rainsford (2011) for details on the discovery of ibuprofen.
26 On language policy and lexicology in Icelandic, see Rögnvaldsson (2022), Albury (2016), Kristinsson (2014) and Kvaran (2004). For Spanish, see Becerro (2020), Azorín Fernández and Sánchez Manzanares (2016), Bellés (2016).
27 See the works of Juan Cuesta y Ckerner, one of the most leading figures in the creation of pharmacy-related vocabularies adapted for Spanish in the 19th century (1868a; 1868b; 1870; 1878a; 1878b; 1883). As Garriga Escribano (2019; 2018), Garriga Escribano and Pérez Pascual (2016) have pointed out, the compilations of dictionaries, even if at slower rate, also participated in the promulgation of scientific terminology.
28 Historically, lexical purism can be traced to the earliest works of translation from the 16th and 17th centuries, in which authors discussed the importance of keeping Icelandic free of foreign influence, followed by a linguistic protectionism sentiment in the 18th and 19th centuries (Ottósson, 1990). See Vikør (2010) and Ottóson (1990) on the purist approaches in scientific terminology.
29 In some ways, the levels of "[s]tandardization and the suppression of variation in favor of a unified norm have always been intimately linked to the conservation of the language in Iceland [...] It would, however, be short-sighted to see the motivation for the strict language planning policy only in the attempt to create linguistic uniformity imposed 'from above' and to educate the speakers in terms of linguistic discipline. For Icelanders, the standard is the vehicle of a literary tradition that documents their cultural roots from the settlement period until the present day. The language is based on an oral tradition that dates to the settlement period and even before and it has been passed on through different generations with only little changes (mostly on the phonological level). As a result, even the youngest generation of speakers today can read and understand the old Sagas from the tenth and eleventh centuries" (Haselow, 2012, p. 147).

30 Major scientific revolutions brought with them obviously their own vocabularies. In the case of pharmacy, the importance of advances in chemistry that took place in France in the 18th century had an impact on French, but also on other European languages. On this, Laszlo (2010) notes that "Lavoisier, lorsqu'il publie en 1789 son Traité élémentaire de chimie, sait l'importance qu'il y a d'asseoir une discipline scientifique sur une « langue bien faite ». En effet, il est alors déjà l'auteur, avec Fourcroy et Guyton de Morveau, d'une Méthode de nomenclature chimique, qui pose les bases de la terminologie encore en vigueur de nos jours. Leur innovation lexicale visait à nommer chaque composé chimique de façon à la fois rationnelle et univoque, contrastant avec la prolifération antérieure des termes usités pour une même substance" [*Lavoisier, when he published his Elementary Treatise on Chemistry in 1789, knew the importance of basing a scientific discipline on a "well-made language". Indeed, he was already the author, with Fourcroy and Guyton de Morveau, of a Method of Chemical Nomenclature, which laid the foundations of the terminology still in force today. Their lexical innovation aimed to name each chemical compound in a rational and unambiguous way, contrasting with the previous proliferation of terms used for the same substance*] (p. 116).

31 See Gordin (2015).

32 See Drummond (1920).

2 Science, Materia Medica and Words

This second chapter delves into the role of language in the development of the fields of chemistry in Iceland and pharmacognosy in Mexico between the late 18th century and early 20th century, showing that linguistic planning has defined the lexicon used and not used in modern pharmacy from its onset. On one hand, focus lies on the coinage of a term in Icelandic that could be used to refer to "chemistry". The science of substances, materials, and changes that paved the way for biomedicine-based pharmacy in its current form. From early on, Icelandic intellectual figures realized the importance of preparing ground for chemistry to emerge and that involved finding the right neologism to describe it. On the other hand, this chapter draws attention to the various attempts to include Indigenous Mexican materia medica in local pharmacopoeias inspired in European counterparts. Reviewing the inclusion of traditional medicinal knowledge, along with its Maya and Nahuatl sounds, provides insights about the managerial aspects of language in the compilation of pharmacy-related literature. Ultimately, in this chapter, I demonstrate that the production of knowledge in major industrial, economic, politic, and intellectual nuclei at the time consolidates in the emergence of linguistic practices. Many of which are a reaction of relatively young nations to the standardization of science as a practice, activity, and systematic study.

2.1 Chemical Sounds

Linguistic planning strategies, at a micro- and macro-level, solidified hand and in hand with the establishment of fields in Iceland in the 18th, 19th, and early 20th centuries. Nonetheless, because the introduction of state-of-the-art scientific discoveries and innovations from Europe and North America was interwoven with print media practices, such as translation. Brokers of knowledge produced abroad played an important role in informing about the latest developments in fields, such as chemistry, medicine, and physics.

DOI: 10.4324/9781003601654-3

Between the late 1770s and the 1790s, Icelanders residing in Copenhagen played a pivotal role in the Icelandic Enlightenment, particularly through the efforts and establishment of Það íslenzka lærdómslistafélag (*The Icelandic Society of Learned Arts*) in 1779 (Sigurðsson, 2010). Translations of foundational academic volumes and textbooks were some of the first venues where Icelandic readers encountered an unprecedented number of new scientific terms, concepts, technicisms, loan words, and neologisms, at times, many densely contained in one single paragraph. As expected, scholars educated abroad were the first to take the initiative of Icelandising (*íslenzka*)[1] entire volumes, despite a several number of challenges (e.g., lack of initiative from the Danish royal authorities). Besides high costs and inconsistent financial support, the absence of an established technical terminology was an obstacle for most of the pioneering translators of scientific volumes. Even before authors engaged with content specificities, in some cases, they were confronted with the lack of a term to name entire fields or subjects, such as chemistry, a foundational disciplinary precursor of pharmacy.

Throughout the 18th and 19th centuries, for instance, besides *epnafræði*, subsequently *efnafræði*, *bræðslufræði*, *samblandsfræði* and *blöndunarfræði* were proposed as viable options to name the rapidly emerging field of chemistry. In 1788, Sveinn Pálsson coined the neologism *bræðslufræði*, or the science of melting, as a translation of "chemien" (p. 96).[2] In the same article, in his discussion about elements, the author furthermore uses the adjective "chemískur" (chemical), to describe the composition-oriented connotation of *Terra elementaris* (Pálsson, 1789, p. 95).[3] This translation suggests that whereas objects in nature can be chemical from an elemental configuration point-of-view, the encompassing scientific field should be best named in terms of the potentiality of those objects being studied. Melting certainly refers to one of the many processes examined and tested in the field in a context dominated by theories about the transformation and states of matter. As Guðmundsson (2019) has observed, Pálsson´s article *Um kalkverkun af jörðu og steinum* (On the calcification of earth and stones), where *bræðslufræði* was first proposed, builds on Georg Ernst Stahl´s ideas; soon to be abandoned once Antoine Lavoisier´s revolutionary work reached Danish, and subsequently, Icelandic scientific circles.

Three decades later, another important figure in the Icelandic cultural scene, Magnús Stephensen coined the neologism *samblandsfræði* (combination science) to best describe "þessa dýrðlegu grein náttúrufræðinnar" [*this glorious branch of natural science*] (Stephensen, 1818, p. 105). Stephensen worked arduously on translations that could serve to divulgate state-of-the-art knowledge among Icelandic readers.[4] He played a special role in promoting some of the most popular ideas that characterized the Danish and German enlightenment,[5] either through translations or articles based on foreign literature.[6] In his contribution *Um járn og stál* (*About iron and steel*), Stephensen dedicates a one-page long footnote on the challenges of finding the right word for chemistry. He starts by stating: "Ég veit ekki að finna í voru máli orð, sem fullkomlega útpryði meiningu orðsins Chemie" [*I cannot find in our language any word that fully expresses the meaning of the word Chemie*] (Stephensen, 1818, p. 105). Leaving his position as mere translator, Stephensen adopts his role as broker and intermediary of knowledge in his detailed explanation about chemistry as a field:

> Hún kennir að þekkja allra hluta undirstöðuefni, að rannsaka þeirra sambland, og að aðskilja þeirra parta í einföld upprunaefni (element). Líka að blanda mörgum þeirra saman aftur til líkrar veru, og ummynda þá í aðrar með þeirri samblöndun. Þar þetta skeður oft án hita eða bræðslu, með samblandi ýmislegra hluta, jafnvel kaldra, eða svo, að ekki megi bræðsla nefnast, finnst mér orðið samblandsfræði næst koma að lýsa eðli þessarar lærdómsgreinar. [*She teaches to know the basic materials of all things, to study their combination, and to separate their parts into simple source materials (elements). Also mixing many of them together again into a similar being, and transforming them into others with that mixture. Since this often happens without heat or melting, with a combination of various things, even cold ones, in such a way that it cannot be called melting, I think the word combination science comes closest to describing the nature of this discipline*]
>
> (p. 105)

Stephensen disagrees with Pálsson, making a case for all those situations in which changes of matter are not necessarily caused either by overheating or melting. Elaborating more in detail on the characteristics and properties of iron and steel gives him also ample opportunity to discuss the intrinsic nature of metals. The discussion furthermore includes some of the earlier translations and established names of the chemical elements known at the time. Guðmundsson (2019) suggests that Stephensen used various foreign manuals as scientific sources for *Um járn og stál*, one of them presumably *Chemisk Haandbog* (1784 (1794)) by Nicolai Tychsen.

However, based on the six examples found in Ritmálssafn Orðabókar Háskólans corpus (Ritmálssafn Corpus henceforth), the neologism *samblandsfræði* does not seem to have been used, except for a short period between 1818 and 1826. A year after Stephensen's article was published, in his dictionary of "rare, exotic and poorly understood words present in Danish books" (*Ordabók, sem inniheldr flest fágiæt, framandi og vandskilinn ord, er verda fyrir í dønskum bókum*), the term is not even mentioned. Gunnlaugur Oddsson's entry on *chymie* reads: "Chymie [...] brædslu- blöndunar-fræði" [*chymie [...] science of melting and mixing*] (Oddsson, 1819, p. 28).[7] The inconsistent use of one term is best exemplified in the inclusion of either Danish, German, Greek, or Latin lemmas, for clarification, often within brackets. Besides the form "chymie" in the previous example, "chemien", "chemie" and "chemia" were also used as referents for the different neologisms being proposed.[8] Today we know none of these names were kept for long and also that the word used for chemistry in modern Icelandic (*efnafræði*) was coined in the context of natural history and philosophy, rather than as an attempt to capture the scientific essence of the field. In 1835, Jónas Hallgrímsson[9], one of the most beloved poets, naturalist, and intellectuals of all times in Iceland wrote:

> Eg skal nú leítast við að gefa lesendum Fjölnis dálítið sýnishorn af jarðfræðinni, eínsog hún lítur út sem stendur, og verður það samt styttra enn skyldi, eínkum vegna þess, að það er mart hèraðlútandi, sem ekki verður fráskírt, nema lesendum væri kunnug heímspeki og helztu greínir úr efnafræðinni*) (Chemie), sem það væri ósanngjarnt að ætlast til, alþýða hefði numið. Mart er það líka, sem enn er óljóst í æfisögu jarðarinnar, því þessi vísindagreín er ennþá mjög í bernsku og fer þessi árin so óðum fram, að valla mun líða lángt um, áður enn hún verði búin að fá annað fegra útlit. [*I will now try to give the readers of Fjölnir a little sample of geology, as it stands at this very moment, and it will still be shorter than it should be, especially because there is much in this regard that will not be completely clear, unless the readers were familiar with philosophy and the most prominent subdivisions of chemistry*) (Chemie),which would be unfair to expect that the general population was familiar with. There is also much which is still unclear about the history of the Earth since this branch of science is still very much in its infancy and time is passing so fast, it will probably not be a long before things start looking much better*]
>
> (Hallgrímsson, 1835, p. 109–110)[10]

The reflection about the role of science that informs the piece *Um eðli og uppruna jarðarinnar* (*About the nature and origin of the Earth*), is best epitomized in the footnote Hallgrímsson signals with an asterisk. There he elaborates on the neologism *efnafræði*, which is a cognate of "efni" (element, material) and "fræði" (science). The note reads that "Í þessari vísindagreín eru ransökuð frumefni náttúrunnar og sameíning þeírra" [*In this branch of science, the elements of nature and their combination are studied*] (Hallgrímsson, 1835, p. 109). As Ringler (1996) and others have pointed out, the article can be read as Hallgrímsson's attempt to familiarize Icelandic readers of *Fjölnir* with the advances in geology at the time, while reconciling Christian faith and empirical approaches to knowledge, moderating the discussion through references to Platonic philosophy. There is nevertheless a sense of tension in *Um eðli og uppruna jarðarinnar*, as it becomes evident that science poses questions and finds answers that do not coincide with religious paradigms. From the various meanings of "efni", the cognate for chemistry appears to derive from "element" in the sense of essence, rather than of a chemical substance. As Guðmundsson (2019) suggests, the neologism is more philosophical than technical in nature, which is not necessarily the approach taken by Hallgrímsson in his life's work of expanding and adapting Icelandic vocabulary in sync with scientific and cultural shifts.

In hindsight, it has been possible to produce a comprehensive list of neologisms coined by Hallgrímsson throughout his lifetime and spamming across various disciplines (e.g., "spendýr" for mammal), aspects of life (e.g., "skipstjóri" for captain), designations for objects (e.g., "stuttbuxur" short pants), and even sense of community (e.g., "vinahópur" for group of friends).[11] Hallgrímsson is considered by many as one of the most salient, influential and impactful figures in terms of contributions to modern Icelandic. As for the word for chemistry, the spelling of Danish and German terms, or adaptations thereof, such as capitalized "Chymie", "Chemía" and "Chemie", kept appearing in many entries throughout the 19th century, until "epnafræði"/ "efnafræði"—Hallgrímsson's proposed neologism—consistently replaced any form used previously.

Important examples of the complete adoption of Hallgrímsson's "efnafræði" include its use in several passages of *Ný félagsrit* (1850) and *Eðlisfræði eptir J. G. Fischer* (1852). In a chapter on the utility and applicability of chemistry for the advancement of agriculture, authors Brynjólfur Snorrason, Gísli Brynjúlfsson, Jón Sigurðsson, Magnús Eiríksson, and Vilhjálmur Finsen write: "Síðan hefir efnafræðin leidt menn til sannfæringar um, að í næring þessari eru bæði loptefni og steinefni" [*And then chemistry has led to the conviction that such a nutrition contains both air elements and minerals*] (*Ný félagsrit*, 1850, p. 105). The use of a definite article (*Ic.* -n) might slightly signal a certain level of familiarization or colloquiality with a relatively new term, or alternatively underline its uniqueness. Within the natural sciences, further important examples of the adoption of "efnafræði" include passages in Magnús Grímsson's *Eðlisfræði eptir J. G. Fischer*, a translation of *Fischers populære Naturlære, til Brug i Skoler og ved Selvunderviisning*, authored by J.G. Fischer and published in 1844. Explaining the scope of chemistry, Grímsson translates the original text as follows: "Efnafræði (Chemie) eða Blöndunarfræði. Hún sýnir hvernig líkamirnir verða aðgreindri í efni sín eða parta" [*Chemistry (Chemie) or mixtures science. She shows how bodies become more differentiated into their substances or parts*] (Grímsson, 1852, p. 2).

Grímsson recurs to three different forms to convey and specify the scientific branch he is referring to. This was not the first written source in which chemistry was being defined as the science of mixing (*blöndunarfræði*), but a relevant one, in the sense that the book was used, for some years at least, in the most prominent educational institution in the country, Reykjavíkurskóli.[12] As one of the oldest examples attests, using "blöndunarfræði", seemed to have been a matter of preference, perhaps choice or understanding of what this relatively new scientific field was about: "Ordatiltæki þetta er tekid af Vísindagrein þeirri, er kallast Chemie, og á Islendsku hefir verid køllud Bløndunarfrædi af sumum" [*This expression is taken from that scientific field called Chemie, one that has been called the science of mixtures by some*] (Ármann á Alþingi I-IV, 1829–1832). From this first appearance, "Bløndunarfræði", eventually "blöndunarfræði", and until it was last documented, the number of entries resembled the ones registered for "samblandsfræði". A difference can be seen between the written sources in which authors and editors adopt and include neologisms without questioning their applicability, and those who discuss their appropriateness or logic. In these two examples, for instance, "efnafræði" is treated as the accepted general term:

> að vera vel að sjer í steinafræði, bergfræði (petrografi) og efnafræði
> [*To know well minerology, petrography and chemistry*]
>
> Yður hefur verið sagt frá því í efnafræðinni, hvers konar þeir hlutir eru, sem eru umhverfis oss
> [*You have been told by chemistry which kinds of things are those that we have in our surroundings*]

In the first example, taken from *Andvari* (1880), no translation in a foreign language within brackets is provided for chemistry, whereas the author is hesitant about the consistency and familiarity of the readers with "bergfræði", a cognate of "berg" (stone) and "fræði" (science). The second example comes from Halldór Kristján Friðriksson's translation of a textbook in physics, *Eðlisfræði eptir Balfour Stewart* (1880).[13] In both cases, the reference to chemistry denotes the acknowledgment and adoption of Hallgrímsson's proposed neologism. Further on the above entries, despite its mission as intellectual magazine on history and cultural matters, *Andvari* presumably had from its foundation in 1874 a different readership in mind than textbooks did, and engaging with the expansion of scientific vocabulary was not its main goal.[14]

In the case of the second part of the entry, it is interesting to note that Friðriksson adapts the book for Icelandic audiences. He adds examples that could be useful for the readers and contributes with some neologisms, although sparingly, compared to other translators. On deviation, for example, Friðriksson adds that "á íslensku mætti það nefna stefnuskekkja eða heimsskauts-afhvarf" [*in Icelandic it could be called directional deviation or polar deviation*] (p. 151). However, on inclination, he observes that "þessi halli segulnálarinnar er nefndur á mörgum útlenzkum tungum *Inclination*" [*this inclination of the magnetic needle is called in many foreign languages Inclination*] (p. 152), without really engaging with the creation of a neologism. On "efnafræði", Friðriksson does not question its use, even in the last part of the textbook, where the original author, Balfourt Stewart, reflects on what chemistry actually studies and how it differs from physics. The decision seems logical. Friðriksson´s translation appeared in a textbook published by the same editorial house that had the year before printed *Efnafræði eftir H.E. Roscoe* (1879), Benedikt Gröndal´s translation of *Chemistry* (1872) by Henry Enfield Roscoe.

Several reasons might explain this. On one hand, the emphasis on *efni* (material/ element), as opposed to processes (e.g., melting and mixing), mirrored new discoveries about chemical elements, their composition, and properties. On the other hand, the amount of translation work that Gröndal (1826–1907) had done on chemistry and possibly his status as a prominent figure simultaneously in the natural sciences and in the literary scene. Gröndal was equally concerned with language and eager to disseminate information about chemistry to Icelandic readers.[15] As it was the case of the new discipline of substances and changes of matter, other names of fields posed a challenge in terms of finding adequate translations. Ideally, neologisms that would retain the intended meaning while averting the opacity of loan words derived from Latin. Para-textually, in *Efnafræði* (1896) Gröndal adds the following remark:

> Önnur eins nöfn og þetta mega til að standa hversu ónóg og vitlaus sem þau eru; "Fysik","Chemi","Botanik","Mineralogi" o.fl. eru einnig í rauninni alvitlaus orð, eins og kemur í ljós, þegar farið er út í þýðingar þeirra á málfræðislegan hátt. [*Other names like these have the tendency to be used if they are inadequate and senseless; "Physics", "Chemistry", "Botanic", "Mineralogy" etc. are also essentially nonsensical words, as it turns out, when one goes into their translations in a grammatical way*]
>
> (p. 4)[16]

From this view, the necessity of transposing the body of knowledge contained in the original book overruled the frustration of using "inadequate and senseless" words. Gröndal was critical about his translations, from a philological perspective, as much as the translations by others, from a scientific perspective. Guðmundsson (2019) notes that he makes quite an unfair remark in his autobiography *Dægradvöl* (1923) about Magnús Grímsson's knowledge on the subject of Eðlisfræði eptir J. G. Fischer: "Magnús þýddi Fischers eðlisfræði með aðstoð Bjarnar Gunnlaugssonar, því sjálfur vissi hann lítið eða ekkert í þeirri grein" [*Magnús translated Fischer's physics with the help of Björn Gunnlaugsson, because he himself knew little or nothing in that subject*] (Gröndal, 1923, p. 194). Although several passages indicate Grímsson, assisted or not, does justice to the original work (Guðmundsson, 2019).[17] Despite the publication of original and translated textbooks by Friðriksson, Grímsson, and Gröndal, in late 19th-century Iceland, chemistry still retained a certain position of novelty. In the January issue of *Landbúnaðarblað* (Agricultural Newspaper) (1899), to a reader's question of whether the journal can explain what the different elements are, the following answer is given:

> Nei, og það af þeirri ofureinföldu, ástæðu að tiltölulega fair lesa nokkuð það, sem að efnafræði lýtur. Efnafræðin er líka ölnbogabarn í skólum vorum, að búnaðarskólunum fráskildum. [*No, and that is due to the very simple reason that relatively few read at all anything related to chemistry. Chemistry is also the hated child in our schools, except for agricultural schools*]
>
> (p. 6)

The authors do proceed to provide some general knowledge of what the chemical elements are and their various types, citing examples and putting them in the context of a readership interested in agriculture. By this time, chemistry on both sides of the Atlantic had expanded exponentially into applications that went far beyond farming. In the case of pharmacy, the introduction of alkaloids, glycosides, and new forms of synthetic drugs[18] had redefined the therapeutic repertoire of the end-of-century materia pharmaceutica. Operationalization of advances and discoveries in biochemistry, physiology, botany, and microbiology progressively resulted in improved medicines. The establishment and growth of new fields and more specialized terminologies had naturally implications at a linguistic level.

Outside scholarly circles, this was particularly noticeable in medicine-related contexts as they largely concerned the general population. Loan words from Danish, German, Greek, and Latin, as well as terms slightly adapted phonetically were for instance common in pharmaceuticals catalogs and "(n)ýyrðasmíð í læknisfræði hófst ekki að marki fyrr en fremur seint" [*coinage of new terms in medicine did not begin until rather late*] (Ottósson, 1990, p. 112). A shift did take place with the appointment of Guðmundur Björnsson (1864—1937) as *Landlæknir* (National doctor) in 1906. In his position as head of the National Health Directorate, much credit is attributed to Björnsson for his achievements in improving the country's health system and policies, especially during difficult times (Ásmundsson, 2014).[19] But also for raising awareness on the use of Icelandic in health-oriented contexts. Like many other figures in the national political scene, Björnsson was a:

> mikill unnandi íslenskrar tungu. Allt sem hann lét frá sér fara í ræðu og riti var á vönduðu, skýru og gjarnan knöppu máli. Hann var snjall nýyrðasmiður og er til dæmis höfundur heitisins "togari" sem kom í stað hins stirðlega orðs "botnvörpungur". [*big enthusiast of the Icelandic language. All his contributions, either in oral or written format, were carefully articulated in a clear and quite concise language. He was a clever creator of neologisms and is, for example, the author of the word "togari" (trawler), which came in instead of the stiff word "bottom projection"*]
>
> (Ásmundsson, 2014, p. 629)

As *Landlæknir* for 25 years and professor at the first medical school opened in the country in 1876, Björnsson advocated the introduction of Icelandic lexicon (neologisms) to replace Danish and German medical terms (Ottósson, 1990). A project that aligned with the lexical purism movement of the time. Consistent in his views, later in his life, Björnsson authored his reflections on Icelandic spelling, phonetic transcription, phonology, and spelling. In *Rangritunarheimska og framburðarforsmán* (Typographical stupidity and pronunciation prepositions) (1912) and *Hljómbætur* (1913) (Sound improvements), Björnsson argues that in some cases spelling should be phonetic, but in some, quite on the contrary, etymology should prime. Whereas it seemed logical to report that his publications on serious diseases resonated with readers widely,[20] less expectingly, his work on phonetics also sparked a debate on phonological spelling, sociolects, and pronunciation.[21] Language was a popular topic of discussion in public discourse, nonetheless because it lay at the core of the independence movement that sought a separation from Denmark. Linguistically, Danish had and retained a special status in Iceland and many generations of intellectuals, scientists, and physicians had completed their education in Denmark. Besides the introduction of qualification protocols for pharmacists adopted in 1828, the opening of Den pharmaceutiske Læreanstalt (*Da.* The Pharmaceutical College) in Copenhagen in 1892 buttressed the recognition of pharmacy as a scientific field at the heart of a regulated occupation.[22] Icelandic pharmacists would be trained under guidance of an authorized pharmacist and sail to Denmark to conduct their examinations (Gunnarsdóttir, 2020).

Latin was a fundamental aspect of the educational curriculum implemented for any aspiring pharmacist in Iceland in the early 20th century, as it was the case for the highest academic tracks and strands in the country at the time. But so was knowledge in chemistry, botany, and pharmacognosy. In *Bráðabirgðareglugerð um nám í lyfjafræði og aðstoðarmannspróf* (Provisional regulation on pharmacy studies and assistant exams), published in 1930, some of the requirements for the oral exam listed included "lestur lyfseðla á latínu og þýðing úr latneskri lyfjaskrá" [*reading prescriptions in Latin and translating them from a Latin pharmacopoeia*] (Björnsson, 1930, p. 5). To obtain a professional certification, applicants for a pharmacist's license were also expected to submit a trainee affiliation card, a lab workbook, and a botanical notebook. The herbarium notebook collection should include 120 types of plants where "Hver tegund jurta sé fest á blað fyrir sig og sé tilgreint latnesskt og íslenzkt heiti jurtarinnar, hvar og hvenær hún er tekin og nafn finnanda" [*Each type of plant is attached to a separate piece of paper along with the Latin and Icelandic name of the plant, specifying where and when it was taken and the name of the finder*] (Björnsson, 1930, p. 4). Establishing, reviewing, and reforming rules for the training of pharmacists in the country was a step towards the standardization of protocols surrounding medicines in Iceland after the turn of the century. The requirements of qualification tests for pharmacists illustrate the change in approaches introduced by the National Health Directorate. Yet, even if Björnsson had vocalized his frustration about the use of loanwords in medical settings (Ottósson, 1990), by 1938, most entries in the register of medicines would still contain a high number of Latin-derived terminology.

Comparatively, pharmaceutical advertisement was equally unresponsive to the adoption of specialized medical language. An illustrative example is the relative absence of terminology updates in Jecorol's advertisements between 1915 and 1930. Produced on both sides of the Atlantic, Jecorol was one of the supplements developed as an alternative to counteract the "Недостача" (*shortage*) (Egereva and Mikhlina, 1929, p. 651) of fish oil,[23] used then widely in the treatment for rickets.[24] The preparation was a "сиропообразная жидкость, сладковатого вкуса, бледно-желтого цвета" (syrupy liquid with a sweet taste and pale-yellow color) (p. 652), produced by a number of manufacturers, being A. Bukowski's pharmaceutical company in Poland one of them.[25] On the origin of the name, Stewart (1894) notes that "*Jecorol* and *panjecorine* are names applied to the extracts of cod liver oil and combined alkaloids already referred to and are derived from the ancient appellation of cod liver oil, *Oleum Jecoris Aselli*" (p. 292). Depending on the manufacturer, Jecorol was also known under the brand names Gaduol and Morrhuol. Placed in the first issue of *Læknablaðið* (Icelandic Medical Journal) in January 1915, the two-line promotional ad of the pharmaceutical preparation read: "Jecorol. Kryddað þorskalýsi með fosforefnum. (Calciumhypofosfit 1 pct. og 1/2 Natriumhypofosfit

Figure 2.1 Advertisement of Jecorol in Ljósmæðrablaðið (*Journal of Midwives Association*), January 1930. Courtesy of Ljósmæðrafélag Íslands.

pct.) Reynist ágætlega við Rachitis og Skrofulose. Börn taka það inn með bestu lyst" (*Jecorol. Seasoned cod liver oil with phosphorus. (Calcium hypophosphite 1 pct. and 1/2 Natrium hypophosphite pct.) It seems to work well to treat Rachitis and Skrofulose. Children take it to their heart's content*)". The terminology in the advertisement remained unchanged for over a decade, repeatedly appearing in the primary journal aimed at physicians and other health professionals without revision.[26] It was only by 1930 that Rachitis (rickets) and Skrofulose (scrofula) were finally replaced with "beinkröm" and "kirtlaveiki" respectively in an adapted version of the ad posted in *Ljósmæðrablaðið* (Journal of Midwives Association), the oldest women's discipline journal in Iceland (Figure 2.1). Conversely, the term for "calcium" remained unchanged.

As for "beinkröm", the term first appeared in four separate volumes of *Rit þess (konunglega) íslenzka Lærdómslistafélags* (Annual of the Icelandic Society of the Learned Arts) between 1789 and 1797, as well as in a few specialized monographs on medicine.[27] By 1851, the entry for Rachitis (Gr. ῥαχίτης rhakhítēs) in *Dönsk orðabók með íslenzkum þýðingum* by Konráð

Gíslason, fixates the medical term with the condition *mollities ossium*. Due to the axiomatic characteristics of several conditions affecting specific demographic groups during a period when tuberculosis was a major health concern, it is not clear how "kirtlaveiki" was always understood. Steffensen (1984) suggests that the preceding term "Skrofulose" might have also referred to cases of *vegetationes adenoides*.[28]

Colloquially, "kirtlaveiki" had several connotations. In 1985, Gunnjóna Sigrún Jensdóttir, a respondent born in 1899, recalling her childhood in Iceland's Westfjords noted that "Sum börn voru alltaf með kolgrænan hor og klístur í augum. Þetta var kölluð kirtlaveiki" (*some children always had coal-green gorse and stickiness in their eyes. This was called glandular disease*).[29] At the same time, she remarks that: "Einn dreng á fyrsta ári tók mamma hennar heim til að lækna af kýlum um allan kroppinn. Þetta var líka kölluð kirtlaveiki" (*One boy in the first year was taken home by his mother to heal from punches all over his body. This was also called glandular disease*) (Sarpur, 1997, n.p.). Reflecting on the pharmacological landscape at the time, Gunnjóna Sigrún Jensdóttir recalls also that her mother (Margrét) "gerði lyf af grösum" (made medicines from plants), such as common butterwort (*Pinguicula vulgaris*) (*Ic.* lyfjagras), wild thyme (*Thymus praecox ssp. arcticus*) (*Ic.* blóðberg), and Icelandic lichen (*Cetraria islandica*) (*Ic.* fjallagrös). Butterwort "var notað í smyrsli" (was used in an ointment), made of fresh unsalted butter in combination with a few other ingredients, which was "gott á bólgur og brunasár" (*good for inflammations and burns*). Lichen extract was used for colds, but she could not recall the exact medicinal application of thyme. At the time, it was "Mjög algengt var að börn væru kvefuð" (*very common for children to have colds*). Not much was done about it, but after "meðul fóru að koma var notuð "saft", sem læknir skaffaði" (*medicines began to arrive, a "syrup" was used, provided by a doctor*). This passage of the oral history of Iceland's Westfjords in early 20th century highlights one aspect of the shift from traditional medicine and the development of chemistry and pharmaceutical preparations (e.g., syrups and calcium-containing preparations). Language logically adapted in sync with the social changes taking place.

At the intersection of science and commerce, the diverse origins of pharmacy-related terminology demanded a level of linguistic adaptation that was challenging to integrate and keep up with. A noteworthy example is the entry for Atroklo fortior in *Skrá yfir lyf úr forskriftasafni lyfsalafélags Íslands* (Register of medicines from the prescription collection of the Icelandic apothecary's guild) (1938):

> Í 1 g. er 0.4 milligröm Methylatropinbromid[30] E 25 centigrömm Bolus alba og 50 centigr. blóðkol. Indicationes: Colitis, obstipatio spastica.

> Intoxicationes intestinales. Spasmolyticum. Dosis: 1 teskeið 1—3sv. á dag. [*For 1 g. there are 0.4 milligrams Methylatropinbromid E 25 centigrams Bolus alba and 50 centigr. Plasma. Indications: Colitis, spastic constipation. Intestinal infectious diseases. Spasms. Dosage: 1 teaspoon 1-3 times a day*]
>
> (Thorsteinsson Scheving and Sæmundsson, 1938, n.p.)

In the compendium, Þorsteinn Thorsteinsson, Scheving, and Jóhann Sæmundsson observe the importance of improving the efficacy and competitiveness of medicines sold to Icelandic patients. The inventory, as the authors explain, aims to inform doctors "jafnharðan tilkynningar um nýjar samsetningar, er nefndin verður sammála um" [*by sending them notifications about new combinations as soon as the committee reaches an agreement on them*] (1938, n.p.). The inclusion of Atroklo fortior is logistic, not linguistic. In what seems to be the case also for many pharma-chemical and pharma-botanical innovations, completing a task efficiently and speedily mattered more than philological concerns.[31]

Illustrated for instance by the entry for a preparation named "Glu-Ca-Fer Trochisci" and which is to be prescribed in cases of "Anæmia hjá börnum og truflanir á calcium-efnaskifti" [*Anemia in children and disorders of calcium metabolism*]; referencing anemia and calcium in Latin and deriving its name from the chemical symbols of elements contained in it. The register of medicines reflects the managerial aspects of the language of pharmacy at best under the entry for "Tablettae D-Vitakalk", where it reads:

> Í hverri töflu er 100 I:E. D-vitamin 35 centigrömm Calciumfosfat og 35 centigrömm Calciumfluconat.
> Indicationes: Rachitis og sjúkdómar með truflum á kalkefnaskifti[32] beina yfirleitt. Urticaria, asthma, spasmophilia, caries dentium, hæmorrhag. Diathesis. [*Each table contains 100 I:E. Vitamin D 35 centigrams Calcium phosphate and 35 centigrams Calcium fluconate. Indications: Rickets and diseases with disorders of bone calcium metabolism in general. Urticaria, asthma, spasmophilia, dental caries, haemorrhage. diathesis*]
>
> (Thorsteinsson Scheving and Sæmundsson, 1938, n.p.)

As the authors explain, the information presented is previously approved by the above-mentioned committee, which is a board put together by request of Lyfsalafélag Íslands (Icelandic Pharmacists Association), and composed of members of guilds of doctors and pharmacists (i.e. Læknafélag Reykjavíkur and Lyfsalafélag Íslands). Besides the development of improved pharmaceutical compounds processed locally, the committee is attentive to the importance of being

> samkeppnisfær um **gæði** og **verð** við útlend verksmiðjulyf, sem hér eru mikið notuð, og hefir jafnframt verið leitast við að endurbæta samsetningarnar á ýmsan hátt, til þess að auka lækningagildi lyfjanna sem mest (emphasis in original) [*competitive in terms of* ***quality*** *and* ***price*** *with foreign factory drugs, which are widely used here, and efforts have also been made to improve the formulations in various ways, in order to increase the therapeutic value of the drugs as much as possible*] (n.p.)

Using bold font to emphasize quality and price seems logical as Thorsteinsson Scheving and Sæmundsson draw attention to the committee's willingness of collaborating with doctors in meeting the needs of patients better, as they explain in the preface. From this angle, the language of pharmacy can be considered as the result of executive decisions aiming to deliver medicines to those who need them. The urgency of registering "Tablettae D-Vitakalk" primed over other considerations and the case of vitamins, amply discussed in the first section of Chapter 3, illustrates this point.

The formation, establishment, and adoption of "efnafræði" as the consistent term to designate the field of chemistry resembled the process by which many words were incorporated into Icelandic in the 19th and early 20th centuries. Influential figures educated, most commonly in Denmark, some of whom resided in Copenhagen for many years, actively engaged in the expansion and adoption of terminology that would reflect state-of-the-art innovations in science. The impact of prominent intellectuals in the establishment of Icelandic neologisms as reactions to chemical innovations in pharmacy bears resemblances to the significance of the works on pharmacognosy authored by Mexican doctors. In both cases, language plays a role in the managed moderation, articulation, and dissemination of knowledge.

2.2 Pharmacognosy and Local Knowledge

A series of entries in Nahuatl, Maya, and Otomi, among other Indigenous languages, in 19th-century and early 20th-century Mexican pharmacopoeias are indexical of scientific, scholarly, colonizing, political and commercial processes.[33] More importantly, as I show in this chapter, these entries are also evidence of the degree to which linguistic planning can be decisive in the transfer, (co)production and (co)dissemination of medicinal knowledge. Focusing on *Ensayo para la Materia Médica Mexicana* by Academia Médico-Quirúrgica (1832)[34] and the third edition of the *Farmacopea Mexicana*, published in 1896, I review the presence of Indigenous local knowledges in the Mexican pharmacological landscape and health-sphere at a brief, but important moment in the history of the country. The main argument is that compiling specimens and medicinal botanical knowledge unavoidably prompted the compilation of linguistic strategies to refer to them.[35] Pharmacopoeias, at times, described as written materializations of colonial projects, even if accidentally and fragmentedly, nevertheless shed light on the origin and silenced sources of the epistemologies inscribed in some of their entries. Simultaneously, judging from the eventual disappearance of Indigenous terms, these compendia help us to understand the logistic dynamics of the language of pharmacy from a chronological perspective.

Awareness of pharmacological qualities found in nature constitutes a compendium of information that has been preserved until present day in many societies. An epistemic transfer from generation to generation that has most likely depended on trial and error, but also, on language. That is because it is difficult to imagine instructing and informing others about medicines without certain levels of descriptiveness and detail, or at least partial use of verbal communication. Certainly, applicability of nature in medicine is still today entwined with oral tradition in various corners of the globe. A knowledge defined by Johann Schmidt as pharmacognosy,[36] or the art/science of knowing how to "Erkennen der Arzeneykörper nach ihren sinnlichen Eigenschaften" [*Identifying medicinal substances by their sensory properties*] (1811, p. 8).[37] An important difference is that, at its peak in the 19th century, *Warenkunde* (Commodity science) as it was practiced by powerful European nations (e.g., Germany), involved processes of examination and quality control of raw material used in medicinal preparations (e.g., authenticity of barks). Specimen acquisition endeavors triggered circuits of mobilization of plants, animals, and minerals across Western labs, museums, universities, factories, and private cabinets. Much of which has left linguistic traces on labels and catalogs, but also in scholarly works, such as pharmacognosy textbooks, pharmacists' training material, materia medica inventories, administrative documents, advertisements in mass media, etc. Yet, even if unintended, these sources provide unvaluable samples of the pharmacy parlance at the time of their production. From the perspective of the European colonization of knowledge,[38] in serendipitous ways perhaps, the recognition of local traditional knowledge can also be regarded as monuments to the epistemic legacy and linguistic heritage of voices suppressed by time and space.

Time is crucial to contextualize the role of oral and written information about the pharmacological properties of natural substances for human health. In the case of the tradition of compiling knowledge on the use of materia medica in books, throughout the West, De Vos (2010) shows quantitatively that "Dioscorides *De materia medica* was unparalleled in providing the basis for the Mediterranean/ European pharmacopoeia" and that "as late as 1865, his materia medica remained the core of the western pharmacopoeia", transcending "historical context, including changes in medical practices and theories" (p. 47). In the case of the Iberian peninsula, as early as 1510, permission is granted to a guild of apothecaries in Barcelona to publish the second oldest pharmacopoeia, *Concordia Apothecariurum Barchinonensium* (1911) (Gómez Saavedra *et al.*, 2016), written in Latin.[39] The first pharmacopoeia written in Spanish was published in Seville in 1527 under the Latin title *Modus faciendi cum ordine medicandi*, influencing the ones published in Nuremberg (*Pharmakopeia Augustiana*) in 1573, Valencia (*Officina Medicamentorum*)[40] in 1601 and London (*Pharmacopeia Londinensis*) in 1618. *Pharmacopeia Londinensis* is especially remarkable as it introduced a style eventually emulated and kept by other pharmacopoeias for almost two centuries. In terms of regulatory measurements applicable to medicines, *Farmacopea Matritensis* was the first Spanish pharmacopoeia to be legally enforced in 1739 (Gómez Saavedra *et al.*, 2016). Linguistically, this was a moment in which a fluctuation of terminologies from other parts of Europe materialized in the almost unrestricted adoption of neologisms and loanwords.[41]

Meanwhile, in New Spain, appearance of words derived from Indigenous languages in the fields of medicine and pharmacy overlapped with medicine-related neologisms (de Enterría *et al.*, 2016).[42] As Fernández Rodríguez (2023) asserts, the eighteenth century holds a special place in the scientific exploration of America, and the focus on language illustrates the scale and impact of the epistemic endeavors and interests of that era. There were important ties among the scientific communities on both sides of the Atlantic,[43] nonetheless because of the tight, but tense, political, economic, and cultural links.[44] Part of the tension could be felt in figures, such as José Antonio Alzate y Ramírez (1737–1799), a priest, scientist, historian, and journalist, fully conversant in the Spanish and New Spanish cultures, attitudes, and approaches to knowledge. Alzate became a prominent actor in the divulgation of information about botany, zoology, minerology, history, and medicine practices of the citizens of New Spain.[45] Furthermore, he was among the first to report on the medicinal practices he witnessed, invariably using words borrowed from the Indigenous languages spoken locally. Alzate's main means of diffusion were the issues of the periodical *Gazeta de Literatura de México* (1788–1795) and the two volumes of *Gazeta de literatura de México* (1790–1794), considered by many, as the first popular science periodicals of its kind.[46] A close reading of his work reveals his disinterest in preserving the institutionalization, standardization, and hierarchization of scientific knowledge. Quite the opposite, the "concern to record and defend Mexican practices leads Alzate to consider the fact that valuable knowledge is not necessarily attained only through officially recognized institutions" (Clark, 2009, p. 174). In fact, he engaged eagerly in discussions about the role of bottom-up knowledge production within large epistemic regimes.[47] In line with his views, Alzate observes:

> Los progresos de la cirugía tan solamente se consiguen en virtud de hechos prácticos: un indio del barrio de Santa María (Curato de Cuernavaca), de oficio cortador de madera, con la hacha se rajó una canilla; de esto le resultó una gangrena; por lo que fué perdiendo la pierna poco á poco, ya el estrago se verificaba en el muslo, separada la rodilla, cuando un indio curandero lo libertó de una Muerte muy próxima, tan solamente con aplicarle enpolvo el Tlalpopolotl un poco tostado, El Tlalpopolotl es lo que los arrieros conocen por liga (á causa de que los Indios cazan pájaros untando varitas que colocan en los sitios correspondientes); con ella curan las mataduras ó llagas que se forman en los lomos de las bestias empleadas en cargar. Es una raíz tuberosa que, se cria con abundancia en la Sierras vecinas á esta Ciudad. ¿A que usos no podrá aplicarse en beneficio de la humanidad? [*The progress of surgery is achieved solely through practical events: an Indigenous man from the Santa María neighborhood (Parish of Cuernavaca), a woodcutter by trade, split his shin with an axe; this resulted in gangrene, causing his leg to deteriorate little by little. The destruction had already spread to his thigh, with the knee severed, when an Indigenous healer saved him from an imminent death by simply applying powdered Tlalpopolotl, slightly toasted. Tlalpopolotl is what muleteers refer to as liga (because Indigenous people use it to catch birds by smearing it on sticks placed in strategic spots); they use it to treat sores or wounds that form on the backs of pack animals. It is a tuberous root that grows abundantly in the mountains near this city. What other uses might it have for the benefit of humanity?*]
>
> (Alzate y Ramírez, 1788)

The passage is short, but resembles many of the pieces Alzate authored throughout his lifetime on the pharmacological properties of plants used by native communities in New Spain. A couple of paragraphs, eventually cited and listed in the *Ensayo para la Materia Médica Mexicana*, a materia medica "pequeño *essay*" [short essay] (III), composed by the members of the Academia Médico-Quirúrgica and published in 1832.[48] An important project carried out under the direction of pharmacist Antonio de Cal y Bracho, largely based on Vicente Cervantes' *Ensayo a la materia Médica vegetal de México*,[49] published in Mexico City in 1791. Some additions in de Cal and Bracho's work includes "Tlalpopolotl" (Nahuatl) (*Flourencia thurifera*), which appears under the entry reserved for "Raiz del manso. (*Helianthus Glutinosus*. F. M. I.)" (p. 36).[50] This is an important first appearance for "Tlalpopolotl" as the information—including the Indigenous denominations—contained in *Ensayo para la Materia Médica Mexicana*, would eventually inform a number of subsequent pharmacopoeias.[51] Perhaps, inadvertently, Alzate had contributed to the inclusion of a Nahuatl term in contrast with the Spanish vulgarism (as the authors referred to that type of term) "Raiz del manso".[52] As the description reads, the plant's

> raíz es balsámica, pectoral, y muy vulneraria. Esta raíz es la que los indígenas conocen con el nombre de liga, y la usan machacada para afirmar las articulaciones, después de haber reducido los huesos disloga dos; para acelerar las cicatrización en las heridas, y mundificar las úlceras [*root is balsamic, pectoral, and highly vulnerary. This is the root that the Indigenous people know as liga, and they use it, crushed, to strengthen joints after reducing dislocated bones, to accelerate wound healing, and to cleanse ulcers*]
>
> (p. 37)

Other names, mentioned in *Ensayo para la Materia Médica Mexicana*, are "Tlacopatli" and "Quauhtotollanensi". An inclusion that reveals the political sentiment and national amour propre behind the drafting of the book and embodied in the promotion of the knowledge of "antiguos mexicanos" (ancient Mexicans).[53]

Linguistic strategies in the short proto-pharmacopoeia align with the view that the language of pharmacy should not only be composed of Spanish terms. As the authors explain in the preface of the book, *Ensayo para la Materia Médica Mexicana* is intended to facilitate the process by which the newly independent First Mexican Republic can

> reemplazar ventajosamente con los indígenos, aquellos que nos conducen los estrangeros, pues á veces llegan á nuestras manos evaporados por su mala repocision, y no pocas adulterados, cuyos inconvenientes se evitarían colectándolos en el país; siendo fácil de este modo reponerlos mejor, tenerlos mas frescos, y á precios mas cómodos [*advantageously replace those brought to us by foreigners with Indigenous ones, as they sometimes reach us deteriorated due to poor storage, and not infrequently adulterated. These issues could be avoided by collecting them within the country, making it easier to store them better, keep them fresher, and offer them at more affordable prices*]
>
> (V)

As the members of the academy explain in their rationale, the repertoire of medicinal plants is patrimony and part of the biocultural legacy of the nation. From this perspective, the inclusion of Indigenous botanical terms, phonetically notated, such "xoxocoyolli", "xolonacatl", "pochotl", "moictle" seems a logical step towards the nationalization of pharmacological knowledge for the benefit of the young republic.

The inclusion of binomials in Latin and the references to various nomenclature systems fixate and legitimize the scientific nature of the book. Paradoxically, the names of the Amerindian languages from which the terms are derived are, with few exceptions, not mentioned. The phonetic notation of words transcribed from Amerindian languages is not always coherent or consistent either, but non-Spanish names were not a novelty either in the 18th century. As De Vos (2007) asserts, since the 17th century, apothecaries "at the Royal Pharmacy, for example, were accustomed to receiving regular shipments of quinine, sugar (used to preserve medicines), cascarilla, calaguala, canchalagua, ipecacuana, Jalap root, Mechoacán root, sassafrass, liquidambar, and sarsaparilla" (p. 216). This means that by the time *Ensayo para la Materia Médica Mexicana* was published in 1832 in New Spain, calaguala (i.e., an extract obtained from *Phlebodium decumanum*) and canchalagua (*Schkuhria pinnata*), for instance, were already part of the pharmacological repertoire of the Spanish Empire. These were Indigenous additions, not brought from Mexico, that had been incorporated in the pharmacists' parlance at the time, showing the link between the logistics of medicines and the logistics of language.

By the time the first edition of *Farmacopea Mexicana* emerged in 1846, the pharmacognosy repertoire was largely influenced by local traditional knowledge, eventually transferred, with some changes, onto the second edition, published in 1874 (*Nueva Farmacopea Mexicana*).[54] The third edition, published in 1896, was the first one to be legally enforced nationwide.[55] The three versions are heavily influenced by the above-mentioned materia medica compendia, but also by the scholarship carried out by the members of the Sociedad Médica de México (Medical Society of Mexico). Their papers "provide scientific endorsement of Mexican flora for therapeutic purposes" (Vega y Ortega, 2016, p. 4) and legitimize the recognition of local traditional knowledge. Examples include August Tourraine's experiment with "yerba del pollo" (*Tradescantia erecta*) to stop episodes of heavy bleeding caused by the extraction of teeth and chemical analysis of "yoloxóchitl" completed by Gumesindo Mendoza and Alfonso Herrera. The influence of traditional botanical knowledge serves as source of inspiration in both cases. Tourraine (1866) notes for instance that "Ayant entendu parler dans le public des vertus hémostatiques de la Yerba del Pollo, je lus l'article d'Alzate et je me desidai à répéter *sur des lapins*, les expériences qu'il avait faites" [*Having heard in the public about the hemostatic virtues of Yerba del Pollo, I read Alzate's article and decided to repeat on rabbits the experiments he had made*] (p. 182). As for "yoloxóchitl", Mendoza and Herrera sought to "verify its authenticity as a popular remedy" (Vega y Ortega, 2016, p. 11). The specificities of the Indigenous origin do not seem important, "La flor, dice el célebre Hernandez, era estimada por los antiguos mexicanos, tanto por sus propiedades medicinales, como por su hermosura y fragrancia" [*The flower, says the famous Hernandez, was esteemed by the ancient Mexicans, as much for its medicinal properties, as for its beauty and fragrance*] (Mendoza and Herrera, 1866, p. 223). Presumably because the vegetal kingdom is inscribed in a telluric national project. By which, the medicinal qualities of plants grown throughout the Mexican topography underlined the qualities of the land and had the merits to be added in the official pharmacopoeias.

The third edition of *Farmacopea Mexicana* also introduces more standardized formats to present information and new linguistic strategies to the formulation of statements and the presentation of pharmacy-related terminology. On "cihopatli"/ "zoapatle" (*Montanoa tomentosa*),[56] the authors note:

> El Profesor P. García Peña encontró además aceite esencial y un principio cristalizado que le pareció ser un alacaloide (4). El Profesor F. Río de la Loza cree que la planta contiene un alcaloide (1893); y el Profesor Quiroz (5) dice haber aislado el ácido orgánico descubierto por el Profesor Federico Altamirano, y le desigma con el nombre de ácido montanoico.(6). [*Professor P. García Peña also found essential oil and a crystalline substance that he believed to be an alkaloid (4). Professor F. Río de la Loza suggested that the plant contains an alkaloid (1893), and Professor Quiroz (5) claimed to have isolated the organic acid discovered by Professor Federico Altamirano, naming it montanoic acid* (6)].
>
> (Herrera *et al.*, 1896, p. 167)

The addition of names of professors, physicians, and scientists is enriched by the incorporation of footnotes and references to contextualize and legitimize the properties listed under each entry. The above-mentioned endorsements of "cihopatli"/"zoapatle" are footnoted respectively: (4) "Estudio sobre el Zoapatle". Tesis para el examen profesional. Mexico, 1888". [*Study on the Zoapatle. Thesis for professional examination. Mexico, 1888*] (5) "Revista Farmacéutica' Febrero, 1804". [*Pharmaceutical Journal. February 1804*] and (6) "Datos para la Materia Médica Mexicana" [*Data for the Mexican Materia Medica*] (p. 167). Interestingly, as footnote 5 shows, some of the information already existed by the time the previous edition was published in 1874, but not included. In terms of properties of "cihopatli"/"zoapatle", not much has been changed since it was mentioned by Francisco Hernández in the 16th century and in *Ensayo a la materia Médica vegetal de México* in the 18th century.[57] Experiments with the plant, such as those conducted by Ramón Alfaro in 1866 are not mentioned,[58] even though his findings confirm the "common" medicinal use of "cihopatli"/"zoapatle": "El jugo de esta planta, en dosis de 60 á 90 gramos, se usa por el vulgo para pro- vocar las contracciones de la matriz: al cocimiento concentrado de las hojas y tallos le atribuyen la misma propiedad" [*The juice of this plant, in doses of 60 to 90 grams, is used by the common people to provoke contractions of the womb: the concentrated decoction of the leaves and stems is attributed the same property*] (p. 168). Focus lies on the use of each medicine listed and on omitting "las substancias que han parecido inertes ó superfluas" [*substances that have seemed inert or superfluous*] (Herrera *et al.*, 1896, XVII). As the authors explain:

> En el largo período de 28 años que ha transcurrido desde la publicación de la primera Farmacopea Mexicana, se han introducido en la Farmacia multitud de preparaciones nuevas, que la Comisión ha debido recoger en la nueva Farmacopea, así como ha desechado otras que se han abolido en la práctica y son de origen empírico: por eso se han suprimido, entre otras cosas, los ungüentos de Agripa, de Artánita y de Sebos; los emplastos Diabotano y de Mucílagos; el colirio de Lanfrac; algunos electuarios poco usados y de utilidad dudosa, como la Triaca, el Absorbente, etc., y se han simplificado, conforme á la ciencia, otras preparaciones, como los emplastos de Cimbrón, de Estabillo, el Divino, etc. [*In the long period of 28 years since the publication of the first Mexican Pharmacopeia, numerous new preparations have been introduced into pharmacy, which the Commission has included in the new Pharmacopeia. At the same time, it has discarded others that have fallen out of practice and are of empirical origin. For this reason, items such as Agrippa's ointments, Artanita's ointments, and Sebaceous ointments have been removed, along with Diabotano and Mucilage plasters, Lanfrac's eye drops, and some rarely used and questionable electuaries like Theriaca, Absorbent, etc. Additionally, other preparations have been simplified in accordance with scientific principles, such as Cimbrón, Estabillo, and Divino plasters (poultices), etc.*]
>
> (XVII)

And yet, the 1896 pharmacopoeia also retains medicines that were not easily attainable, or that have been included in European counterparts, despite their rareness or unavailability, for instance, plants growing in places far away from Mexico. As discussed in the introduction, an example thereof is the entry for Icelandic moss (*Centraria islandica*), retained in pharmacopoeias, despite being less and less prescribed by Mexican physicians. In many ways, it is possible to see this as a sign of reluctance to question or differ from older and more established written works. After all, *Farmacopea Mexicana* needed to reflect the projects of internationalization of science and modernization set in motion by Porfirio Diaz's presidency (1876–1880 and 1884–1911), commonly known as the Porfiriato period.[59]

An attitude the authors endorse. Best characterized by their explanation about linguistic choices throughout the book:

> Los nombres que en los idiomas francés é inglés tienen las substancias empleadas en terapéutica, ha creído la Comisión deberlos incluir en la

> sinonimia, tanto por haberse generalizado el uso de estas lenguas en México, donde se consultan por los sabios los obras escritas en esos idiomas [*The names of substances used in therapeutics in French and English have been included in the synonyms section by the Commission, both because the use of these languages has become widespread in Mexico and because works written in these languages are consulted by scholars*]

but also because "algunos médicos (aunque indebidamente) acostumbran poner sus recetas en los ya mencionados idiomas" [*some doctors (albeit improperly) are accustomed to putting their recipes in the aforementioned languages*] (1896, XIV). There is a rationale behind the order of languages and names, as the authors proceed to explain. In the first place, they include "1°, el nombre con que es más generalmente conocida" [1°, *the name by which it is more generally known*], followed by "2°, los otros nombres científicos ó vulgares que aún lleva ó ha llevado en otras épocas de la ciencia" [2°, *the other scientific or common names that it still bears or has borne in science in other time periods*]. In third place, each entry includes the "nombre técnico latino" [*Latin technical name*] and finally "4°, el que tiene en las Farmacopeas francesa é inglesa" [*the one under which they appear in the French and English Pharmacopoeias*] (1896, XVII). Importantly, as the reference book contains several preparations that are "meramente nacionales y desusadas en Europa" [*purely national and not used in Europe*], it has been impossible to offer synonym terms in French (XVII), which is neither the case for those preparations that "que no se usan en Inglaterra ni en los Estados Unidos, ó que aunque llevan los mismos nombres, son distintas por su composición" [*are not used in England or in the United States, or that although they bear the same names, they are different by their composition*] (1896, XVII).[60]

Para-textually, the authors articulate in the preface the role they assign to language in the presentation of the information contained in *Farmacopea Mexicana* from various perspectives. Firstly, there is a desire to facilitate the process by which the book can be accessible to foreign readers. It mattered to include scientific terms that could help users identify the plant-, animal- and mineral-based substances listed in the work. Secondly, linguistic opacity certainly worried the authors, nonetheless because there was the risk that the scientific legitimacy of the pharmacopoeia could be called into question. Although the order in which the languages presumably does not aim to be hierarchical, it provides insights about the intricate relation between science and linguistic choices, authoritativeness, and legitimacy. Thirdly, the decision of prioritizing on the "sinonimia" (synonymy) of the names resulted, even if unintended, in the inclusion of a wider number of terms in Indigenous languages. Many of these "nombres vulgares" (vulgar names) and their *synonyms* were in fact translations, at times provided in combination with geographical details. For instance, the entry for "Calancapatle de México", reads: "Palancapatli; Lechuguilla; Guaje (en Michoacán) [in Michoacán]; Nanahuapatli", preceded by the "*Solidago mexicana, L., et S. velutina*, D. C" (p. 37). As opposed to the entry for "suet": "Sebo.—Suif, Franc; Suet, Ing.; *Sebum ovillum*" (p. 147) with a translation in French (Franc) and English (Ing), but without any name in Indigenous languages. As the introductory text explains:

> Fácilmente se comprende la importancia que tienen para nosotros los nombres de las drogas en los diversos idiomas que hablan los naturales del país, ya por facilitar al farmacéutico la adquisición de los productos indígenas, ya para las investigaciones que más adelante puedan emprender el médico ó el naturalista, ó ya, en fin, para los estudios históricos ó filológicos. [*It is easy to understand the importance that the names of drugs in the various languages spoken by the natives of the country hold for us, whether to facilitate the pharmacist's acquisition of Indigenous products, to aid future research by doctors or naturalists, or, finally, for historical or philological studies*]
>
> (1896, XIII)

At its peak as a discipline, the focus of pharmacognosy lied on the quality, availability, and identification of materials in nature with pharmacological properties (Cahlíková *et al.*, 2020). In this sense, the linguistic choices in *Farmacopea Mexicana* responded to the needs of identifying substances by using the names locals would recognize. The logistics of obtaining the right materia medica would depend on the linguistic logistics that informed the drafting of the book. In the process, the members of the permanent commission in charge of compiling the pharmacopoeia were in the position to model the language of pharmacy in late 19th century. Each materia medica entry operated as a label in Spanish or "en los idiomas mexicano, otomí y otros de los que se hablan en nuestro territorio" [*in the Mexican language, Otomi and other that are spoken in our territory*], as well as French and English for "los más conocidos" [*the most known ones*] (XIII). In this sense, those interested in the pharmacopoeia as a source to carry out, as mentioned by the authors, philological studies, would need to start by unraveling the origin, provenance, etymologic and phonetic notation of the hundreds of Indigenous words of the Mexican language of pharmacy at the turn of the century.

In 19th-century Mexico, as it was the case in Iceland in the 18th and 19th century, language plays an important role in understanding the context in which scientific knowledge is communicated and shared, as well as in embracing it. The dissemination of pharmacy-related knowledge took place through the management of language. An aspect that confirms the similarities between science-driven linguistic approaches and task-oriented logistic operations. Icelanders needed to embrace, accommodate, learn, and appropriate chemistry as a field, as much as Mexicans needed to adopt and adapt compilation protocols of pharmacological knowledge. In both cases, it mattered to find or design the right word(s).

Notes

1 The verb "að íslenzka" meant to translate a book into Icelandic and is commonly found on the covers of scientific monographs in the 19th century and early 20th century. Often, the names of the translators served also as reference of their scholarly authority, status within academic circles, notoriety as public figures or excellent command of Icelandic and the foreign language in question. In modern Icelandic, the notion of "að íslenska", with a slightly different spelling, is still used mainly in the context of poems, special literary pieces or as a means to emphasize the authorship of a particular translator or writer.

2 Sveinn Pálsson (1762–1840) made important contributions in the form of articles about natural science and medicine, besides travel books with detailed descriptions about Icelandic flora (Steindórsson, 1981).

3 In contrast, for instance with *Terra inflammabilis*, *Terra calcarean*, and *Terra metallica*.

4 Magnús Stephensen (1762–1833) was a lawyer, governor, and founder of Landsuppfræðingarfélagið, a cultural association established to disseminate knowledge and information about the latest innovations in arts and sciences in Europe.

5 Senner (2000).

6 An important aspect to consider is that "it can be maintained that in the period under review ideas connected with the Enlightenment and ideas connected with romanticism intertwined to form one general current in the ideology of many Icelanders and that this was reflected in their nationalistic attitudes" (Sigurðsson, 2010, p. 381).

7 Neither of the words is to be found in the hand-written dictionary *Íslensk-latnesk orðabók* compiled by Jón Ólafsson (Bjarnadóttir, 2019).

8 Almost all the etymological forms of *kemi* listed in the seminal *Ordbog over det danske Sprog* (1926), can be traceable in Icelandic translations of academic monographs or compiled dictionaries at some point during the 18th and 19th centuries.

9 Jónas Hallgrímsson (1807–1845) played an important role in transforming the literary scene through his own work, but also contributed to the political emancipation of Iceland, for instance through the establishment of the journal *Fjölnir*. He also engaged with the divulgation of information and scholarly work produced in Denmark while remaining tightly associated with the Icelandic Independence Movement.

10 For a translation of the first third of the article in English, see Ringler (1996).

11 See Þráinsdóttir and Einarsdóttir (2022) for a comprehensive list.

12 Except for some details, a book largely based on Johann Heinrich Helmuth's *Volks-Naturlehre*, originally published in 1786 under the name *Volksnaturlehre zur Dämpfung des Aberglaubens* (Guðmundsson, 2019). The popularity of the book in Germany and Denmark probably had an impact on its reception as a textbook beyond its original intended audience.

13 Balfour Stewart was a Balfourt Sewart (1828–1887) was a Scottish prominent physicist, meteorologist, member or the Royal Society and active figure in the divulgation of knowledge among specialized and wide audiences through primers, textbooks, and academic monographs.

14 Cultural magazine established in Copenhagen in 1874 by Hið íslenska þjóðvinafélag, and except for1878 and 1892, ever since published yearly.

15 Gröndal was passionate about the importance of educating the public in Iceland and mediating knowledge. In a letter, for instance, dated 8th February 1880, he wrote Willard Fiske, Icelandic language professor and librarian at recently founded Cornell University, he says: "Þér sögðuð við mig að jeg skyldi segja yður frá ef mig vantaði nokkuð – en það sem mig vantar og lángar til að eiga, er Microscop sem stækkar hér um bil 3-400, jeg sé opt þessi Microscop nefnd í blöðunum, svo sem frá Leiß í Jena, en jeg hef hvergi séð hvað þau kosti, en jeg er líka hræddur um að þau sé of dýr fyrir mig. Það er samt ómissandi fyrir mig að hafa þettað instrument, og minni stækkun (magnifying power) vil jeg ekki hafa. Menn vita ekkert um hin mikroskopisku dýr hér á Íslandi, af því náttúrufræðingar dvelja hér aldrei nema stutta stund og gefa sig þá við öðru. Þér sjáið náttúrlega í útlöndum öll blöð og vitið hvað allir hlutir kosta, og ef það ekki tæki allt of lángan tíma, þá ímynda jeg mér að þér munið gjöra svo vel og segja mér eitthvað um þetta" [*You told me that I should tell you if I needed anything - but what I need and would like to have is a Microscop that expands here by about 3-400, I often see this Microscop mentioned in the papers, such as from Leiß in Jena, but I have not seen anywhere what they cost, but I am also afraid that they are too expensive for me. It is still essential for me to have this instrument, and I don't want less magnification (magnifying power). People know nothing about the microscopic animals here in Iceland, which is why naturalists never stay here for more than a short time and then turn to other things. You of course see all the papers abroad and know how much everything costs, and if it doesn't take too long, I imagine you will be so kind as to tell me something about this*], see Bragadóttir (2017, p. 106).

16 Another similar exampled found in the same book: "Aptur vandræða-orð! því öll efnafræði er líka bygð á öflum. Í vandræðum köllum vèr "mekaniska Fysik" "aflfræði"; þar á móti er "kemisk Fysik" ekki "efnafræði"[*Again trouble-words! because all chemistry is also based on forces. Problematically, we call "mechanical physics" "mechanics"; on the other hand, "chemical physics" is not "chemistry"*] (Gröndal, 1886, p 4)

17 Guðmundsson (2019) shows the importance of textbooks in the establishments of natural sicences in Iceland (Link: https://uni.hi.is/einar/).

18 See Cowen and Helfand (1990).

19 Björnsson faced sanitary crises caused by tuberculosis, Spanish flu, influenza and lepra.

20 Björnsson wrote for instance *Um berklasótt* (About Tuberculosis) in 1898.
21 See Guðmundsdóttir (2022) for a discussion of the impact of Guðmundur Björnsson's work on phonetic spelling within the context of the history of approaches to pronunciation in Iceland.
22 Bjerg Larsen *et al.* (2004) observe that the shift from an absolute monarchy to a constitutional once, based on democratic participation, had an impact on how the welfare state regarded "pharmacy and the way that pharmacy policies were elaborated" (p. 13), resulting eventually in more strict regulations at all levels, but also more engagement and recognition of pharmacy and its role in society.
23 Based on pathophysiological understanding of rickets in early 20th century, Alfred Hess, along with others, pioneered the use cod liver oil supplementation to treat the condition in an African American community in 1917 (See Hess and Unger, 1917).
24 The efficacy of Jecorol was questionable, if the right amount and density of cod oil was not used (Stewart, 1894). After carrying out lab experiments on rats, Egereva and Mikhlina (1929) concluded that, "Jecorol же не оказывает ни профилактического, ни лечебного действия при экспериментальном рахите. Он не содержит противорахитическото фактора и не может быть рекомендован для леченая рахита" [*Jecorol has neither a prophylactic nor a therapeutic effect in experimental rickets. It does not contain anti-rachitic factor and cannot be recommended for the treatment of rickets*] (p. 653). The experiment was conducted on Jecorol produced in the Soviet Union by Tatmedtorg. A pharmaceutical company which only operated for four years, between 1923 and 1927, and eventually changed its name to Tataptekoupravlenie (Stepanishchev, 2020).
25 See Muralis *et al.* (2023).
26 The advertisement kept appearing in this format in the medical journal *Læknablaðið* until 1926.
27 See Sigurðsson (2010).
28 On Skrofulose in Iceland in the 19th century, see Finsen (1874) and Steffensen (1984).
29 In an interview conducted by SHÞ in July 1985.
30 In many ways, this phonetic transcription resembles contemporary phono-semantic matching, the main approach applied to names of pharmaceuticals in modern Icelandic.
31 On the profound effect of science on language, see for instance Degaetano-Ortlieb and Teich (2018) on the case of English as result of the evolution of modern science from the mid-17th to late-19th century, resulting in linguistic diversification and specialization, but also grammatical conventionalization.
32 An example of creating "new lexemes by using native material" (Haselow 2012, p. 145).
33 See Figueroa Saavedra and Berzaín Varela (2023) on the connection between translation from Nahuatl, bilingualism, and health communication in 18th-century New Spain.
34 See Academia Médico-Quirúrgica (1832), *Ensayo para la materia medica mexicana.*

35 In the 21st-century, advocacy for the rights and certain forms of agency of plants and animals calls into question the anthropocentric view of using nature mono-directionally for the benefit of humans. It is not clear whether this view also holds for cases of survival or disease, or predominantly as a call for action and reaction to the environmental crisis triggered mainly by the implementation of profit-driven economic systems where nature is mainly considered as a source of raw materials.

36 The term is coined from the Greek words *φάρμακον* (pharmakon: drug) and a νῶσις (gnosis: knowledge).

37 According to the definition in *Lehrbuch der materia medica*: "Das wäre denn die eigentliche Naturhistorie der Arzeneykörper (*Pharmacographia*), oder in wie fern es dabey auf ein Erkennen der Arzeneykörper nach ihren sinnlichen Eigenschaften ankömmt, die Arzeneyenkunde (*Pharmacognosis*)" [*That would then be the actual natural history of the medicinal bodies (Pharmacographia), or to what extent it depends on recognizing the medicinal bodies according to their sensory properties, the medicine knowledge (Pharmacognosis)*] (Schmidt, 1811, p. 8). In present-day German, Arzeney (Medicine, remedy, drug of all kinds) is also spelled "Arznei", "ärznei", "erznei", "ärzenei" (Frühneuhochdeutsches Wörterbuch, 2023).

38 See Boumediane (2016).

39 The first official pharmacopoeia of which we have records is the *Receptario Florentino* (*Nuovo receptario composto dal famosissimo Chollegio degli eximii Dottori della Arte et Medicina della ínclita ciptá di Firenze*), published in 1498 in Florence (Gómez Saavedra, 2016).

40 From an intra-linguistic point of view, compared to other similar editions and previous pharmacological compilations, Spanish pharmacopoeias "exhibit a higher degree of consistency", perhaps a sign of "a general trend of increasing standardization over time" (De Vos, 2010, p. 43).

41 The important translation of French chemical nomenclature in 1788 was, for instance, quickly adopted in educational settings in Spain (Bello, 2010).

42 One example thereof is the "vocabulario desarrollado durante la implantación de la vacuna, observamos en cambio la ausencia de voces fluctuantes, la presencia de neologismos de rápida difusión (vacuna, vacunar o vacunación) y la misma coincidencia con las voces empleadas en la península advertida en el vocabulario de las fiebres epidémicas" [*the vocabulary developed during the implantation of the vaccine, we observed instead the absence of fluctuating words, the presence of rapidly spreading neologisms (vaccine, vaccinate or vaccination) and the same coincidence with the words used in the peninsula noted in the vocabulary of epidemic fevers*] (de Enterría *et al.*, 2016, p. 89).

43 See Crawford and Gabriel (2019), Bleichmar (2015).

44 The exchanges also included materia medica. De Vos (2007) calculates that at least "700 different types of medicinal plants and plant products arrived in Spain throughout the 1770s and 1780s, including for instance, the Guaco herb, considered to be a 'universal antidote' for all snakebites" (p. 25).

45 For more on Alzate y Ramírez, see: García (2017); Cruz León (2008); Terán Elizondo (2004); Saladino García (1990); Peset (1987).

46 In terms of format (e.g., paratextual framework), Alzate's *Gazeta* "se convierte por lo tanto en modelo o ejemplo de construcción de la escritura científica puesta en página y de su representación en ésta, para otros escritores en su momento y editores de este tipo de publicaciones que surgirían después" [*thus becomes a model or example of the construction of scientific writing in print and its representation therein, serving as a reference for other writers of the time and for editors of similar publications that would emerge later*] (Valdez Garza, 2022, p. 185).

47 An example thereof is Alzate's opposition to Carl Linnaeus botanical nomenclature, and which sparked fiction between him and Vicente Cervantes, editor of *Gazeta de Literatura*. See Perez Ramos (2022) for an insightful reflection on the relation between scientific translation related to pharmacy and paratextual notes, specifically in the case of a female translator in the early 19th century.

48 In certain way related to its predecessor, *Tablas Botánicas para el más pronto y fácil Estudio de la Ciencia* (1825) by Julián Cervantes and probably influenced by Pedro Calderón's translation of *Posologie*, originally written in French by Cadet de Gassicourt (1829).

49 Schifter Aceves (2014) notes that 113 out of the180 plants described in *Ensayo para la Materia Médica Mexicana* are to be found in Cervantes' *Ensayo a la materia Médica vegetal de México*. The preface also alludes to the importance of the information collected by Francisco Hernández de Toledo (1514–1587), posthumously summarized, and edited in *Quatro libros de la naturaleza y virtudes de las plantas y animales que estan receuidos en el uso de la medicina en la Nueva España*.

50 On its use, in the previous edition of the pharmacopeia, Herrera *et al.* (1884) notes that, commonly, "goza de gran reputación como vulneraria: su conocimiento se emplea contra las disenterias. Con el jugo se preparan aparatos inamovibles" [*it enjoys a great reputation as a vulnerary: its recognized applications include the treatment of dysentery*] (p. 93).

51 Izquierdo (1952) suggests that *Ensayo para la materia medica Mexicana* and other similar works sought mainly to have "Indigenous substances substituted for exotic ones" (p. 60).

52 According to Amador (1899), Yerba del manso was also known as "Compuestas", "Tlapopolotl", "Tlacopatli" in Aztec language and "Guaguat" in Opata (a now presumed extinct language also known as Tegüima, Tehuima, Tehui, Eudeve, Eudeva, Heve, Dohema, Jova, Joval, Tonichi, Sonori and Ure and Teguima in Opata).

53 References to "antiguos mexicanos" in masculine form are found on pages 6, 8, 24, 25, 27, 54. The only reference to "antiguas mexicanas" (ancient female Mexicans) appears under the entry reserved for "Pepitas de Sopilote. (*Swietenia Mahagoni, L.--Tzopilotltzontecomatl, ó Tzopilotl, Hern*" (p. 32) in reference to a fruit that had been used as a cosmetic in pre-Columbian times.

54 "Vega y Ortega (2016) asserts that "almost all the physicians and pharmacists who published the 12 papers in the GMM [Gaceta Médica de México], the physicians Luis Hidalgo y Carpio, Maximino Río de la Loza and Lauro María Jiménez, and pharmacists Alfonso Herrera and Gumesindo Mendoza,

participated in the *Nueva farmacopea Mexicana*" (Vega y Ortega, 2016, p. 15).

55 See Saldaña Azuela (1994), Hersch Martínez (2001), Aceves Pastrana (2007), Morales Sarabia and Aceves Pastrana (2011), Morales Cosme and Viesca Treviño (2016), Reynoso *et al.* (2017), Govantes Morales (2021).

56 Gallegos (1983) asserts that "Zoapatle, the common name for Montanoa tomentosa (Cerv.) s.s.p. tomentosa (Funk.), is a well-known plant in Mexico. Its name was given by Vicente Cervantes in honor of the outstanding Mexican physician of the 18th century, Luis Montaña" (p. 212).

57 Properties also previously also described by Fray Bernardino de Sahagún in *Historia General de las Cosas de la Nueva España* (*Florentine Codex*) (1540–1585) and Martin de la Cruz in *Libellus & medicinalibus indorum herbis* (1552) (Gallegos, 1983).

58 Alfaro "mentioned that the properties of cihopatli extract were being investigated scientifically and that it could probably be used in the future to treat feminine maladies" (Vega y Ortega, 2016, p. 11).

59 On the Porfiriato, Godínez Reséndiz and Aceves Pastrana (2014) note that the "principales actividades científicas y de investigación se llevaron a cabo durante el periodo histórico conocido como el Porfiriato (1876–1911), donde el gobierno permitió que el país estuviera en constante contacto con la ciencia europea y con los científicos originarios de ese continente" [*The main scientific and research activities were carried out during the historical period known as the Porfiriato (1876–1911), where the government allowed the country to be in constant contact with European science and with scientists from that continent*] (p. 56).

60 In terms of structure and division (natural products, chemical products, preparations, instructions), the third edition of *Farmacopea Mexicana* resembles the fifth edition of *Farmocopea Hispana* (1865). A book that took almost 50 years to complete due to the complexities of Ferdinand VII's reign (Gómez Saavedra *et al.*, 2016).

3 Morphology and Pharmaceuticals

This chapter reviews how pharmacy instrumentalizes language to produce a lexicon that aligns with its needs in terms of reference and conventions. Focus lies on the fact that the history of pharmacy,[1] at the nexus of natural sciences (i.e., chemistry and biomedicine) and commerce, results in the amalgamation of fixed naming standards and commercially driven and tactic branding projects. As the first part shows, the names used by millions today have long histories, which can be traced to knowledge transfers and remote places from the surgery room and pharmacy cabinet where they are kept today. Along this line of thought, the second part proceeds with a reflection of how medicinal knowledge exchanges materialize in words used to organize pharmacy-related and -derived taxonomies. Looking back to important milestones in the development of medicines, for instance, in the context of the Transatlantic Exchange, although mainly focusing on early 20th century, this section addresses the marks and markers of mercantilization in the establishment of drugs as references in the therapeutic collective imaginary. From this perspective, I suggest that acknowledging the input of local traditional and Indigenous knowledges in the discoveries (epistemic containers) and development of medicines underlines the fact that pharmacy in many forms and shapes existed long before pharmaceutical companies became industrial emporia. Reviewing the place of botany in the advancement of pharmacy,[2] I insist on the difference between pharmacy as knowledge and pharmaceuticals as objects, defined nominally (epistemic labels) by logistic conventions. At last, the chapter closes with a discussion that centers on the sociolinguistic identity of medicines and the implications thereof in spoken language, contextualized by a link between the history of names and their adoption as terms.

3.1 Infixes, Identities and References

Morphologically, much of the scientific lexicon is built on units, carefully put together to provide information at various levels, and as I show

DOI: 10.4324/9781003601654-4

in the next paragraphs, a morpheme, as small as an infix provides cues about the history and epistemic systems linked to its origin. The linguistic presence of quinine (cinchona),[3] for instance, registers the Quechua ancestry of a therapeutic practice, as much as it chronicles and contextualizes the journey of the bark across European medical cabinets and laboratories. The lingering presence of **-quin-** as an infix in present-day chemistry serves as a reminder of a knowledge transfer from Andean medical repertoire, ergo, from an ancestral knowledge system.[4] Using quinaquina,[5] as a febrifugal obtained from the Cinchona genus trees,[6] throughout the Andean region, preceded protocols of experimentation to which the bark was later submitted in European and North American laboratories.[7] Though not in exactly the same settings, trial and error and observation most likely also took place across the Andes, before a bark would receive the name "bark of bark"/"holy bark", that is "quinaquina".[8]

The movement of the bark across imperial networks had a linguistic component to it that has survived for centuries. The Quechua legacy of the molecular composition of quinine [(9R)-6-methoxy-8α-cinchonan-9-ol], in its various nominal versions (e.g., quinaquina, quina, kina), has a place in modern-day chemical nomenclature. More importantly, it has retained its place in the knowledge system where it first emerged. Focusing specifically on **-quin-** as an infix, for instance, in chloro**quin**e, Atova**qu**one (quone > quinine + one), or Ato**quin**ol (allyl phenylcinchoninester) (Figure 3.1) an opaquer name, is not about deconstructing terminologies a la Saussure or Derrida. It is about magnifying, underlining, and highlighting the ways in which linguistic evidence can be useful in contextualizing the creation of the language of pharmacy.

In this sense, language serves as a marker of the epistemic identity of raw materials used for medicinal purposes. By epistemic identity, I refer to the processes, sites, and actors involved in the knowledge production behind therapeutic and pharmacological innovations, such as Cardiazol-Quinine (Figure 3.2), commercialized in France in the first half of the 20th century.[9] Regardless of the settings in which they emerge, language plays a role in any strategy devised to pass knowledge on to others. Health-related vocabularies are found in all languages, confirming that no society is devoid of a medicinal repertoire. Applying knowledge to restore human health is equally patrimony of biomedicine (Western science), as much as of traditional epistemic systems, such as the one in which quinine first emerged to treat fevers.

In many ways, the Quechua- derived infix **-quin-** found in present-day chemical nominal system is historically, epistemically, linguistically, and culturally relevant, nonetheless because it is still part of present-day

Figure 3.1 Advertisement of Atoquinol (allyl phenylcinchoninester), produced by Ciba Laboratories (1900/1950). Author: Georges Goury.

Source: Bibliothèques de Nancy, P-FG-ES-00795. Courtesy of Galeries Limédia.

therapeutic repertoire, as the following excerpts from the Corpus del Español NOW show:

1 Acerca de la posible resistencia de el parásito infectante, se debe administrar quinina o quinidina. Actualmente se recomiendan las terapias combinadas con artemisina (Venezuela 2013).
[*Regarding possible resistance of the infecting parasite, quinine, or quinidine should be administered. Combination therapies with artemisinin are currently recommended*]

2 Con la generalización de el uso de la mepacrina y el desarrollo de la cloroquina y de la primera quinina sintética (Spain 2019).
[*With the generalization of the use of mepacrine and the development of chloroquine and the first synthetic quinine*]

Figure 3.2 Advertisement of Cardiazol-Quinine (1900/1950), produced by Cruet Laboratories in Paris. Author: H. Biais. Courtesy of Galeries Limédia. Public domain.

3 La mayoría de los casos de malaria, como lo hizo la resistencia a la cloroquina en la década de 1980, contribuyendo a millones de muertes », explican (Spain 2019).
[*The majority of malaria cases, as resistance to chloroquine did in the 1980s, contributing to millions of deaths", they explain*]

In its own way, it also engenders the Andean locality and eventual isolation and synthetization of a molecule, based on an Indigenous febrifugal. I suggest that **-quin-** encapsulates the bioprospecting, colonizing and exploitative aspects of cinchona's history, as much as the authority of Andean medicinal knowledge. In this sense, language registers the apparently opaque traces and impact of quinaquina on medicine. Mass media outlets offer insightful examples of how this interplay materializes in written language, for instance through the imperceptibility of the Quechuas infix **-quin-** in Spanish-speaking news:

4 Reconstituyente elaborado a base de extractos de malta, nuez de kola, coca, quinidina y de los fosfogliceratos propios de la sangre (Colombia 2018).
[*Restorative made from extracts of malt, kola nut, coca, quinidine, and phosphoglycerates commonly found in blood*]

5 Petición de medicamentos antipalúdicos como cloroquina, quinina, clindamicina, primaquina, artesunato, quinidina, mefloctina, entre otros que son necesarios para contrarrestar los diferentes tipos de malaria (Venezuela 2018).
[*Request for antimalarial drugs such as chloroquine, quinine, clindamycin, primaquine, artesunate, quinidine, mefloquine, among others that are necessary to counteract the different types of malaria*]

6 Diversos alcaloides, de los cuales los más abundantes e importantes son: quinina, quinidina, cinconina y cinconidina. Aparte, posee también principios astringentes (Spain 2014).
[*Various alkaloids, of which the most abundant and important are: Quinine, quinidine, cinchonine, and cinchonidine. Besides, it also has astringent principles*]

7 Venezuela realiza el pago para adquirir primaquina y cloroquina (para el tratamiento antimalárico), solicitado a el laboratorio BSN Medical en Colombia (Chile 2019).
[*Venezuela makes the payment to acquire primaquine and chloroquine (for antimalarial treatment), requested from the BSN Medical laboratory in Colombia*]

From a linguistic logistic perspective, the managerial dimension of moving a bark across apothecaries' cabinets is not very different from the planning operation of adopting and standardizing an infix within the dominant internationally recognized chemical language. Even if minimally, **-quin-** is a token of Andean knowledge heritage, and simultaneously, evidence of processes of linguistic coordination and task planning, involving producers, brokers, translators, and communicators of science and knowledge. Although the spectrum of medicines to treat malaria is broader today, understanding quinine marked the onset of antimalarial treatments. In similar ways as the synthetization of curare and cocaine prompted new horizons in the development of medicines used as anesthetics and paralyzing agents in surgical procedures. The line of medicaments, eventually developed based on cocaine and curare, both derived from Indigenous medicinal repertoires in South America, are still identifiable in medicines containing the infixes **-cain-** and **-curo-** in their scientific names.[10]

Figure 3.3 Curare darts and quiver Musée d'ethnographie de Genève, Item ETHAM 007525.

Source: Photographer: Rama. Creative Commons Attribution 3.0.

But before atracurium, cisatracurium, pancuronium, rocuronium, and vecuronium could become important muscle relaxant precursors in the clinical repertoire of options,[11] the adoption of a body of knowledge preceded their eventual development.[12] The infix **-curo-** is an adapted sound of the local knowledge of the Guyana Mukusi word "wurari" (Burr and Leung 2014). Possibly, a phonetic notation version of "ourare", "ourari", "urare", "woorari" and "woorara", among others;[13] all Indigenous names used to refer to the bush rope plant (also known as devil doer) (*Strychnos toxifera*). Curare refers specifically to the poison that can be extracted from the plant and that was originally used in arrows (Figure 3.3) as part of a hunting technique in the Amazon Basin (Lee, 2005), consisting of immobilizing an animal to facilitate its capture.[14]

The first European source on this practice was the extensive letters by Pietro Martire d'Anghiera, published in early 16th century, followed by Sir Walter Raleigh's descriptions in 1596 (Figure 3.4).

In 1834, on curare, "Humboldt claimed that the chief ingredient was a strychnos species of the Loganiaceae. He named the plant 'strychnos toxifera', which was the chief constituent of calabash curare" (Evans,

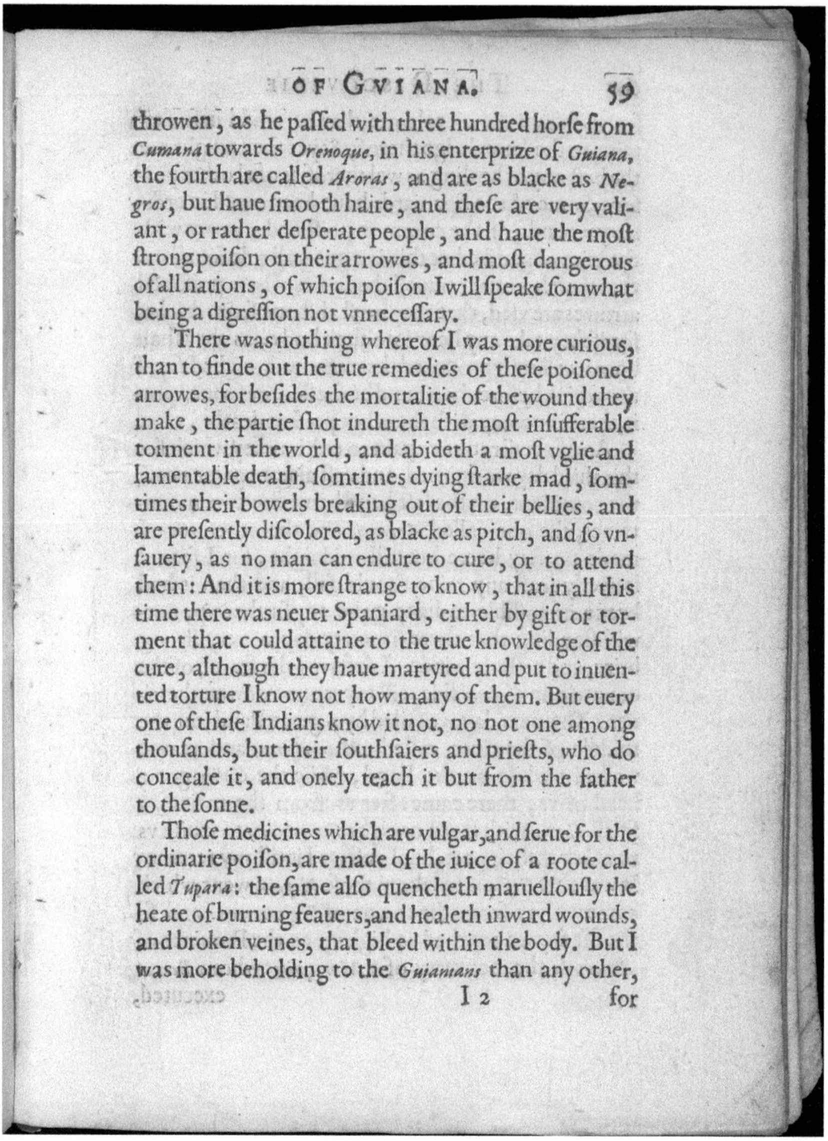

OF GVIANA. 59

throwen, as he paſſed with three hundred horſe from *Cumana* towards *Orenoque*, in his enterprize of *Guiana*, the fourth are called *Aroras*, and are as blacke as *Negros*, but haue ſmooth haire, and theſe are very valiant, or rather deſperate people, and haue the moſt ſtrong poiſon on their arrowes, and moſt dangerous of all nations, of which poiſon I will ſpeake ſomwhat being a digreſſion not vnneceſſary.

There was nothing whereof I was more curious, than to finde out the true remedies of theſe poiſoned arrowes, for beſides the mortalitie of the wound they make, the partie ſhot indureth the moſt inſufferable torment in the world, and abideth a moſt vglie and lamentable death, ſomtimes dying ſtarke mad, ſomtimes their bowels breaking out of their bellies, and are preſently diſcolored, as blacke as pitch, and ſo vnſauery, as no man can endure to cure, or to attend them: And it is more ſtrange to know, that in all this time there was neuer Spaniard, either by gift or torment that could attaine to the true knowledge of the cure, although they haue martyred and put to inuented torture I know not how many of them. But euery one of theſe Indians know it not, no not one among thouſands, but their ſouthſaiers and prieſts, who do conceale it, and onely teach it but from the father to the ſonne.

Thoſe medicines which are vulgar, and ſerue for the ordinarie poiſon, are made of the iuice of a roote called *Tupara*: the ſame alſo quencheth maruellouſly the heate of burning feauers, and healeth inward wounds, and broken veines, that bleed within the body. But I was more beholding to the *Guianians* than any other,

I 2 for

Figure 3.4 The discoverie of the large rich, and bewtifvl empire of Gviana by Sir Walter Raleigh (1596), page 59, describing the poison "curare". Courtesy of Wellcome Collection gallery (2018-04-03).

1948, p. 737). The most salient figure on the subject, from a pharmacy perspective, was Claude Bernard, a Paris-based professor in experimental physiology, who discusses the possible implications and potential applicability of curare-based preparations in *Leçons sur les effets des substances toxiques et medicamenteuses*, published in 1857. Curare was first used for general anesthesia in a Leipzig-based hospital in 1912. In doing so, Arthur Läwen's administration of the substance opened a new chapter in the history of medicine. The eventual synthetization of tubocurarine in 1935, the main ingredient of tube curare extracted from *Chondrodendron tomentosum*, made it possible to introduce the paralyzing agent in systematic ways in surgery rooms in Canada in 1942. In 1948, Evans wrote, "Curare is not an anaesthetic, but its use has revolutionised anesthesia" (p. 736).

As for the local anesthetic cocaine, before its introduction as a pharmacological innovation, in the aftermath of the European colonization of the Americas, coca became a curiosity inscribed in the pages of explorers and scholars intrigued by its biocultural and medicinal use among Indigenous peoples. Thus, "local anaesthetic use can be traced as far back as the highlands of ancient Peru, where the Incas used the leaves of the coca plant as a local anaesthetic to relieve fatigue and hunger and to produce a feeling of exhilaration" (Gies *et al.*, 2017, p. 176). Quite an important detail bearing in mind that "local anaesthesia forms the backbone of pain control techniques in oral surgery" (p. 176). Cocaine was first isolated by Albert Nieman in 1859 and eventually introduced by Karl Koller as a local anesthetic agent for cornea surgery and Edmund Jellinek for larynx interventions (Jácome Roca, 2003). Cocaine was for a long time an ingredient in the frizzy drink Vin Mariani, until its pharmacological effect was fully understood. After its isolation in 1860, cocaine was first used as an aesthetic precursor in 1884 (Gies *et al.*, 2017, p. 176). A milestone was reached when Nils Löfgren and Bengt Lundqvist first synthesized Xylocaine,[15] the precursor of the anesthetic Lidocaine in 1943. The first amino amide-type local anesthetic agent in history and "has become the drug of choice in most clinical settings because of its potency, rapid onset, and lesser allergenicity" (p. 176). The substances were named after "xylo" (tree) and "lido" (ligno > wood) in combination with **-caine** (alluding to cocaine).[16]

The name xylocaine was created as a union of its most important raw product, xylidine,[17] and the general suffix use for local anesthetics, "-caine" (Gordh *et al.*, 2010, p. 1435). The original name of lidocaine was LL30, the abbreviated combined initials of Löfgren and Lundqvist (Wildsmith, 2011). Initially, the trade names chosen were "Xylocain" and "Lidocain", but the vowel -e was eventually added in the United States, whereas the Lignocaine became the generic name used in the United Kingdom. The discovery of lidocaine was a significant advance in the development of local anesthetic drugs.[18] Hanson and Hingson (1950) noted that in the "present reports are that ninety per cent of the Swedish dentists are using Xylocaine in preference to procaine today" (Gruber, 1950, p. 141). The generation of medicines made possible by the work on lidocaine is ever since identifiable by the Quechua-derived infix **-ca-** in combination with the affix **-ine.** It includes substances, such as mepivacaine and bupivacaine, synthesized in 1957, prilocaine, synthesized in 1969 and etidocaine, synthesized in 1972 and currently, "the pharmaceutical industry continues to explore the development of safer and more effective local anaesthetics in a pursuit that has come a long way since the earliest experiments with cocaine" (Calatayud and González, 2003). Cocaine, curare, and quinine were part of local pharmacological traditions long before the encounter between Indigenous populations and European explorers. The phonetic adaptations of their names in Amerindian languages eventually were kept and adopted in scientific taxonomies and consequently in subsequent discoveries. It is impossible to predict for how long **-ca-**, **-cur-** and **-quin-** will retain their place in pharmacy, but their mere presence in 21st century nominal conventions illustrate to which extent language is logistically managed and planned.

3.2 Pharmaceutically Nominal and Nominally Pharmaceutical

Pharmacy can be present in the body, as much as in our speech, public discourse, and daily communication. There is knowledge behind the creation of medicines, in whatever format they may come, either derived from traditional knowledge, science, or other forms of knowledge. Medicines, stemming from biomedicine, have names, often written in leaflets and packages, spoken by those using them, strengthening in the process their linguistic and epistemic presence in everyday life. Each one of them is different. The names of those health-restoring and -preserving artifacts contain markers of history, heritage, knowledge systems, (forced) mercantile exchanges, (abusive) commercial practices, (racial) social hierarchies, scientific nomenclatures, but also, the sounds of languages, often silenced, and yet present through words, such as quinine, tubocurarine, rapamycin (See chapter 5) and indeed, aspirin.

The sounds of traditional knowledge, seem to have remained encapsulated, perhaps caged, in substances used for their, positive and negative, biological effect on other living organisms. It is well documented that "tabaco" (*Nicotiana*) and "kinakina" (*Cinchona*), returning briefly to one of above-mentioned examples, are among the novelties to which Europeans became acquainted after the encounter between the so-called New and Old Worlds took place. There was certainly a time when tobacco was considered a panacea. Some argued that it could be a useful disinfectant and others claimed it could be an effective substance to resuscitate drowning patients.[19] The possibility that its name derives from Arabic طُبّاق *ṭubbāq* / طُباق *ṭubāq*, or alternatively from طباق áabāq (medicinal herbs) underlines the fact that tobacco was once part of the pharmacological repertoire on both sides of the Atlantic and beyond. Its (dubious) therapeutic applicability has been outweighed by its (hazardous) recreational use. As early as the 17th century, evidence-based concerns emerged about the damaging effects of smoking. Today, there is consensus about its toxicity. While it retains its place as a sacred plant, an entheogen, for some, the most important sacred plant. For good or bad, tobacco has had a place in the global medical parlance for centuries. Be either of Taino or Arabic origin, the *Nicotiana* genus in all its species is an example of the poison-cure paradox that the term *pharmakon* encompasses, as Jacques Derrida famously pointed out:

> Une telle opération ne serait pas possible si le pharmakologos n'abritait en lui-même cette complicité des valeurs contraires, et si le pharmakon en général n'était, avant toute discrimination, ce qui, se donnant pour remède peut (se) corrompre en poison, ou ce qui se donnant pour poison peut s'avérer remède, peut apparaître après coup dans sa vérité de remède. L'« essence » du pharmakon, c'est que n'ayant pas d'essence stable, ni de caractère « propre », il n'est, en aucun sens de ce mot (métaphysique, physique, chimique, alchimique) une substance.
>
> (Derrida, 1972, p. 150)

> [*Such an operation would not be possible if the pharmako-logos did not already harbor within itself that complicity of contrary values, and if the pharmakon in general were not, prior to any distinction-making, that which, presenting itself as a poison, may turn out to be a cure, may retrospectively reveal itself in the truth of its curative power. The "essence" of the pharmakon lies in the way in which, having no stable essence, no "proper" characteristics, it is not, in any sense (metaphysical, physical, chemical, alchemical) of the word, a substance*]
>
> (Johnson, 1981, p. 25)[20]

The comparison Derrida makes between the notion of *pharmakon* as a poison-cure analogy applicable to philosophy, a point originally made by Dioscorides, I suggest, equally applies to linguistics. Pharmacy-related lexicon depends on scientific language as much as science relies on the authoritativeness and standardization of its linguistic articulation. As the examples extracted from corpora in the subsequent chapters show, the presumed rigidness, coherence, consistency, and cohesion of scientific terminology nevertheless allows for ambiguity in terms of sense and meaning. Particularly in regards to the origin of substances, but also of more abstract concepts. As it will become clear, the language of pharmacy is logistically managed in sync with its purpose. This is because it ought to operationalize an interface between producers and brokers of knowledge, health mediators/professionals and patients. But despite the indisputability of, for instance, prefixes, infixes, and suffixes in supporting structurally nominal categories, categorizations, and nomenclatures, they have a history of their own. In their short, almost imperceptible (mostly) monosyllabic constitution, they retain traces that have registered the chronicle of knowledge and science. Perhaps, pharmacy, despite its self-assured connotation inherited from scientific predominance, keeps nevertheless in its own vocabularies, pieces of legacy and traces of cultures and forms of knowing, often perceived as un-, pre-, non-scientific. If the infix **-spir-** in aspirin is a reference, even if indirectly, to the willow bark and myrtle leaves that once inspired the synthetization of acetylsalicylic acid, then perhaps **-spir-** is evidence that the language of pharmacy is also the language of traditional knowledge heritage. Perhaps a reason to alert us about the perils of fetishizing science, as much as the perils of imaging a division that renders medicines too chemical, too evil, too colonizing, and too accomplice in the erasure of seemingly *lost* forms of medicine. Many of them do save lives. But lives are also saved through many other forms of pharmacy that do not emanate from laboratory-factory-distribution commercial networks.

Pharmacology is inherent to human survival. Before the revolution in the development and production of drugs in the 20th and 21st centuries, expertise in animal-, plant- and mineral-based preparations was the most common approach to the use of substances in health- restoring and - preventing settings. Lafuente and Valverde (2004), rightly observe that for administrators and scientists, it is fundamental to believe that "nature can be captured in words, figures, lines, shading, gradients, or flows" (p. 134). Sources that self-assuredly convey the idea that raw materials, procedures, and places can be reduced to written descriptive outlines. And yet, "this reducing of nature to representation is necessary, but it is not enough" (p. 134). There is certainly a biopolitical element behind the preparation and administration of substances. Knowledge about the biological effect of substances on those taking them underpins any attempt to manage disease using agents external to the human (or animal) body. These classifications indeed emerge from establishing canons that can provide a blueprint of nature. Yet, as Frake (2012) observes, "(I)n the context of different cultures the task is to state what one must know in order to categorize objects correctly" (p. 443). Language is central to these (successful and failed) attempts at all levels and in all settings. When pharmacognosy first emerged in the 19th century, as the study of crude drugs based on substances found in nature, the underlying premise was that cure for all ailments were to be found in nature. It was a matter of searching for them. Finding the right substances involved, beyond doubt, reporting on *discoveries* that would invariably become "piece[s] of paper in a filing-cabinet" (p. 134). The pursue for novelties applicable to medical contexts has played a central role in explorations and cataloging projects, as much as in justifying the commodification of nature. But since most parts of the world have been inhabited for thousands of years, searching for medicinal plants involved a type of intercultural and interlinguistic interaction. Much has been written on the mobilization of natural sources for medicinal purposes. None of it, or hardly any of it, can be chronicled without adding the recorder's encounter with an unknown word, a possibly misunderstood label, a misspelled name, a phonetically challenging concept, or an inaccurate translation.

Language, or lack thereof, mediates, modulates and (un)settles the processes by which new pieces of information percolate new venues, contexts, small communities, societies at large and far away continents. The passages in the history of the Cinchona genus are an example thereof. As Segers and Van Molle (2022) assert, knowledge is not passed on across networks statically and when it circulates, it does so most likely through emulation and imitation defined by differentiation and variation. Copying and overtaking the use of a febrifugal used across the Andes is the simplest explanation for the mobilization of the quinine-based preparations. If a categorization is essentially "a device for treating new experience as though it were equivalent to something already familiar" (Frake, 2012, p. 443), Cinchona existed as part of the repertoire of antifever formulas in Europe before it was imported from Peru or was known to a European explorer. Similarly, "a definition of a Navajo plant category is not given by a list of botanical species it contains but by a rule for distinguishing newly encountered specimens of that category from contrasting alternatives" (Frake, 2012, p. 443).[21] After Carl Linnaeus, this unnamed Navajo plant would potentially be confinable to a binomial. A set of words that would render it findable and categorizable within a system detached from the relation the Navajo might have with it. By proxy, in terms of Lafuente and Valverde (2004), a combination of names that would seek to homogenize diversity by transforming peculiarities into information derived from coupling nomenclature and morphology. Yet, returning to the above-mentioned Quechuan quinaquina and the undisclosed Navajo example, language remains outside the biopolitical dimension of classification. At least in the sense of pharmacy exerting power onto existing entities. Pharmacy existed long before pharmaceuticals became materializations of technology, as it is conceptually understood in the 21st century.

Objects and subjects, including plants, have a place in the ontology of those societies aware of their existence. From a Heideggerian perspective, the potentiality of nature to become medicinal anticipated any coordinated scientific project derived from accumulation of knowledge as opposed to a tool for survival. Hence, it is logical to assume that the language of pharmacy also preceded technology and (modern-day approach to) science. In its most basic format, a pharmacological preparation could have been expressed through a hand gesture, an exclamation, or a shout. A system to articulate orally information about remedial and curative preparations is found in all languages. As diverse as it may be, it nevertheless, relies on a distinction between substances. Distinguishing, most likely, at a linguistic level, induces epistemological categories, in whatever format, shape, dimension, association, (dis)order or (dis)array it may be. From this angle, the sounds of traditional pharmacological knowledge do not imply an imposition of reformulated categories and/or engender an erasure of existing entities.[22] And they certainly do not either, in the case of pharmaceuticals derived from biomedical and chemical experimentation. In both cases, the names of medicines are designations and references to the outcome of trial-and-error episodes, practices and recorded emulation, variation, and adaptation.

Medicines are distinct from one another and so are the words or any type of sound or linguistic articulation to refer to them. This type of precision is required as their mere instrumentality depends on the correct identification of their properties. Pharmacological substances engender the information about their efficacy and applicability. They are epistemic objects. If "ciencia no sólo se hace con palabras, sino que se hace dentro de una lengua" [*science is not only done with words, but is done within a language*] (Lafuente and Giménez Toledo, 2020, n.p.), the language of pharmacy is made up of the words needed to establish differences, for instance, among natural and artificial, substances. It is therefore a linguistic system linked to the reportorial interaction between the human body and external agents. In the 21st century, this interplay has exponentially stretched the linguistic presence of medicines in everyday life and public discourse. Medicines have an identity of their own, as much as the users of medicines experience them as determinant parts of their own identity. Suddenly, the name of a pharmaceutical becomes more than a catchy label successfully assessed and approved by regulatory authorities. Some become part of cultural phenomena, such as contraceptives (colloquially known as "the pill") or the pre-exposure prophylaxis treatment for HIV (colloquially and clinically known as "Prep").

Names of drugs bolster the nomenclatures used to distinguish between formulas, but also to create a sociolinguistic imprint of a product. Drug names are certainly outputs of science and technology, but also of marketing and commerce. Their epistemic-linguistic identity is built on strategies, or "linguistic tricks" in Gangwal and Gangwal's terms (2011). Some of which include the use of "plosive letters (P, T, or D) to convey power, or fricative letters (X, F, S or Z) to imply speed" (4). These are choices most likely driven by a medicine's "consonance ou l'idée que leur fabriquant souhaite lui voir véhiculer" [sound or the idea that their manufacturer wishes to convey by it] (Faure, 2014: 76). Nexium, Clarinex, Celebrex, Zinetac, Corex, Xanax, Zyban, and Zithromax are examples of the overrepresentation of "x" in pharmaceuticals names that Gangwal and Gangwal (2011) correlate with the fact that this is one of the letters that "look(s) better in print, make(s) sounds people like saying and are associated with innovation" (p. 4). From a branding perspective, the "flamboyant and swashbuckling" twenty-fourth letter in the Latin alphabet is "associated with science fiction, high tech, computers, and automobiles" (p. 4), and suitably, marketing medicines benefit from these references. Products, pharmaceuticals included, have monetizable identities interlinked to linguistic devices. Above-mentioned Xanax is, for instance, not available in Iceland as a medicine to be prescribed to patients, but it is nevertheless present in many more ways and hence, also in Icelandic:

8 Heimapressað Xanax í dreifingu (2018).
[*Homemade Xanax in circulation*]

9 Xanax! Lyf sem algengt er að ungmenni séu að misnota og fikta með (2018).
[*Xanax! A drug that is commonly abused and experimented with by young people*]

Besides the linguistic aspect, the actual unauthorized mobilization is a matter of concern, best exemplified by the warning issued by Embætti landlæknis (Icelandic Directorate of Health) in 2018:

> Lyf sem eru flutt inn af einstaklingum eða búin til á ólöglegan hátt eru mun hættulegri þar sem þau geta innihaldið ýmiss aukaefni. Dæmi um slíkt lyf er Xanax sem ekki er á markaði hér á landi. Xanax inniheldur alprazolam og iðulega önnur efni sem bætt hefur verið við. [*Medicines imported by individuals or manufactured illegally are much more dangerous as they may contain various additives. An example of such a drug is Xanax, which is not on the market in this country. Xanax contains alprazolam and often other substances that have been added*] (n.p.)

The message requires on behalf of the readers to have absorbed Xanax as a notion, reference, name, or entity, even if vaguely, within their linguistic repertoire. In the case of forbidden substances, the subject of discussion in Chapter 5, the logistic aspect of language shows in the coinage of vocabularies emerging at the periphery of any prescriptive boundary.[23] As other examples from the corpora illustrate, the place of substances in the repertoire of familiar names is defined by many factors. Beyond their therapeutic use, the identity of medicines is also defined by their presence in media outlets, controversies, rareness, and revolutionary properties, as the case of Ozempic, discussed in Chapter 5. Labeling practices for drugs have changed precisely to enhance their recognizability. References to precursors or to the history of medicines are nowadays secondary. At first glance, the origin of, for instance, morphine, quinine, artemisinin, digoxin, vincristine, taxol, ergotamine, pilocarpine, ephedrine, and atropine is opaque. Despite the increased popularity of naturality, greenwashing, and pseudo-environmentalism as marketing strategies, these pharmacological substances, developed originally from traditional knowledge, are not commonly associated with plants. The opaqueness of names in the language of pharmacy is indicative of the processes by which medicines have become almost exclusively disassociated from the history behind their creation. In this sense, the chemical language differs from the pharmacy language. The first one evolves in sync with epistemic additions, whereas the second one is active and reactive in sync with salable applications of knowledge.

As the discussion in the subsequent chapters progresses, the transactional and negotiating role of language in pharmacy-related settings becomes clearer. Discussing the navigation of concepts instrumental in the mobilization of medicines confirms that indeed the "use of taxonomic systems is not confined to librarians and biologists; it is a fundamental principle of human thinking" (Frake, 2012, p. 441). A process facilitated by language at its core and as the basis of the relation between society and knowledge, operating along different dimensions and not necessarily co-opted by science or allopathic medicine, in the case of pharmakon. Importantly, "a classification is not of itself an explanation" (Bowker and Star, 2000, p. 319). Considering the suffix **-ol**, for example, in the coinage of Geniol, one of the most popular analgesics and antipyretics in Argentina could potentially offer new insights about clustering and categorizing in the language of pharmacy. Within nomenclatural protocols, **-ol** is reserved for certain preparations,[24] but in the case of Geniol, more than a classificatory role, **-ol** seems to be instrumental in aligning a drug with a specific time and place. Historian Daniel Balmaceda, perhaps nostalgically recalls that:

> Además, agregamos, contaba con el tan característico final en "ol" de aquellas primeras décadas del siglo XX en que ya existían el tónico anticaspa para el pelo Javol, talco para niños Vasenol, jabón para pelo Pilol, jabón dentífrico Odol, el quitamanchas Limpiol, el antiséptico Aniodol, el colorante para teñir el pelo Aureol y el jabón a base de plantas aromáticas Plantol, entre otros. [*Additionally, we add, it featured the characteristic "ol" ending typical of those early decades of the 20th century, when products such as the anti-dandruff hair tonic Javol, the baby powder Vasenol, the hair soap Pilol, the toothpaste soap Odol, the stain remover Limpiol, the antiseptic Aniodol, the hair dye Aureol, and the aromatic plant-based soap Plantol, among others, were already available*]
>
> (2020, n.p.)

In this scenario, Geniol appears to have entered the therapeutic imaginary of Argentineans as a member of an intangible cluster of similar, yet different, health-, body- and home-related consumables, all beneficial for the general wellbeing. An idealized genealogy that may explain its ubiquity demonstrable through examples found in the various corpora:

10 Una pequeña radio. Te ofrece el diario, agua fresca, caramelo y hasta genioles. En los últimos tiempos la atención se fue perdiendo (Argentina 2017).
[*A small radio. He sells newspaper, fresh water, caramel and even genioles. In recent times, attention has gradually been lost*]

11 De hacer lo cuando lo rodeaba un escuadrón de pibes remendados que lo llenaban de genioles y estampitas, y él los escuchaba y sonreía y metía los manones (Argentina 2013).
[*To do so when he was surrounded by a squad of ragged kids who showered him with Geniol tablets and holy cards, and he would listen to them, smile, and shove his big hands*]

Geniol exists morphologically in a pluralized form, but also in the diminutive form "geniolito", later adopted as Geniolito, with capital G, and revamped as a Geniol-based formula targeting the needs and taste of small children. Although informally, Balmaceda reflects on a possible classification of names based on the termination **-ol**. One that occurs in a parallel dimension to the standardized agglutinating function of the suffix in marking any relation to alcohol and phenol. It is precisely the appropriation of the language of pharmacy at a societal level the core of the next chapters. A process only possible due to the logistic processes that precede the diffusion, percolation, circulation, and mobilization of pharmacy-related objects in the first place.

3.3 Sociolinguistics of Pharmacy

By talking about medicines, elements of scientific language become part of our linguistic repertoire, a point best illustrated through the discussion about aspirin, Lantus, and Viagra throughout this book.[25] Before Lantus (Insulin glargine), a long-acting form of medical insulin, entered both the repertoire of Icelandic and Spanish patients at the turn of the 21st century, a series of processes took place. Planning, procurement, production, and distribution precede its adoption and use. As much as the scientific, legal, and commercial perspectives are crucial, so are the linguistic ones. Proceeding with the example of Lantus. The term is found 104 times in the News on the Web (NOW) Spanish Corpus,[26] in sentences such as:

12 Mi endócrino me ha cambiado a lantus, porque según esto es "mejor" y el efecto dura más (Spain 2016).
[*My endocrinologist has switched me to Lantus, because apparently, it is "better" and its effect lasts longer*]

13 El más vendido es la insulina Lantus, la cual produce en la planta de el Estado de México (Mexico 2017).
[*The best seller is Lantus insulin, which is produced at the plant in the State of Mexico*]

14 Sanofi está buscando nuevos productos para compensar las caídas de su insulina estelar Lantus (Peru 2018).
[*Sanofi is seeking new products to offset the decline in sales of its flagship insulin, Lantus*]

15 El Lantus de Sanofi aumentó su precio de 99,35 dólares en 2010 (Panama 2018).
[*Sanofi's Lantus increased its price from 99.35 dollars in 2010*]

Lantus emerges in the pharmaceutical landscape as a response to the impending need of managing type I and type II diabetes. By default, it also emerges in the linguistic landscape inhabited by the physicians, pharmacists and more importantly, patients, their families and those around them (health-sphere). Through media outlets, Lantus becomes part of other communicational and conversational settings, as the abovementioned examples show. Long before this was the case, the actual medication, and the words needed to speak about it underwent a series of steps: planning, procurement, production, inventory, and distribution. This could be encapsulated in the next example, extracted from the

Icelandic MÍM Corpus. Here, "insulin" is used to refer to all brands of diabetes-management medication derived from medical insulin:

16 En það að sykursjúkir fari saman í lest yfir landamærin til að kaupa sér insulin er ekki aðeins til að kaupa lyfið (2019).
[*But the fact that diabetics travel together by train across the border to buy insulin is not just about purchasing the medication*].

In daily conversation, Lantus is one of several nouns used to refer to insulin. A term immortalized by Edward Albert Sharpey-Schafer in 1916 and derived from "insula" (*L.* Island or islet).[27] Ever since Leonard Thompson became the first person to receive an injection of insulin in 1922[28], this protein hormone has become part of the biomedical vocabularies of millions around the world. Names of brands have become synonymous with insulin as the corpora consulted for this book show. Yet, medical insulin has retained its place as a discovery that has miraculously saved an uncountable number of lives, but also a term anchored in the therapeutic imaginary and linguistic repertoires of millions. In the Icelandic case, as the next example hints at, the anticipated amount of "insulín" sent from a pharmacy in another town would have probably saved the life of the patient in question:

> Þrátt fyrir viðeigandi mataræði, alkalia og stimulantia ágerðist coma, og dó sjúklingurinn eftir 7 daga legu. Insulín (frá Siglufjarðar apóteki) kom of seint. [*Despite appropriate diet, alkali and stimulants, the coma worsened, and the patient died after 7 days of hospitalization. Insulin (from Siglufjörður pharmacy) arrived too late*]
>
> (Heilbrigðisskýrslur, 1933, p. 43)

The previous example is interesting for the discussion I advance in this book. On one hand, from the perspective of the circulation of medicines and on the other, from the linguistic perspective. The insulin needed for the patient located in the surroundings of the Akureyri municipality did not complete its journey successfully. Ca. 60 kilometers, between the pharmacist and the patient came in the way. From the reports, we do know that both "alkalia" and "stimulantia" were administered. They were physically available to the physician, according to the terminology at the time. Today, neither of them is used in modern Icelandic. An example of the effectiveness of the neologism policy Icelandic biomedical terminology has undergone, even if it structurally took off "fremur seint" (rather late) (Ottósson, 1990, p. 112). Logistics dictates the flow of medicines from production to the hands of physicians and patients, as much as, I suggest, linguistic logistics steers the flow of scientific language and its adoption, dissemination, adaptation, and appropriation. As nouns, both Lantus, and insulin appeared in the language and spread across it as a result of various processes of linguistic planning, procurement, production, and distribution. I suggest that the language of pharmacy emerges from the mobilization of figurative containers and labels navigating across networks of dissemination. Whereas the containers are the sites of knowledge production negotiation (e.g., scientific journals), the labels are the sets of concepts, terms, and names (e.g., vitamins). The networks are the webs (e.g., media outlets) where the points of reception (e.g., users) come in touch with new linguistic elements.

Returning to the example of Lantus, it needed to be coined as a word that referred to a medication approved to be used in the treatment of diabetes. In certain ways, science precedes scientific language. Once discoveries are made, new vocabularies are needed to articulate, describe, and disseminate information about them. At the same time, one could argue that language precedes science in the sense that it is embedded in a solid, infinitely fertile set of structures that allows for the expansion and adoption of terminologies. Sanofi-Aventis, the pharmaceutical behind Lantus, might have had good reasons to choose Lantus as a commercial name. Most likely, the same rationale that resulted in the coinage of, for instance, Toujeo, and Basaglar, both of which are equivalent trade names for the same medication. There were several planning steps behind the baptism (in Kripkean terms) of these brand names. Diabetes-management treatments can be regarded as results of a major scientific breakthrough and embodiment of knowledge about human metabolism. They are also evidence of the application of biomedicine and pharmacy and their role in human health. In like manner, I argue, scientific language is planned and managed. It is conceived, devised, and discussed. The language of pharmacy and biomedicine involves the procurement of words, particularly nouns, to label, categorize, describe, and explain pharmaceuticals. As the above-mentioned examples taken from the Icelandic and Spanish corpora show, the presence of pharmacy in our daily lives is such that scientific terms and trade names have become part of our conversations for over a century. Even beyond, but I mainly focus on the 20th century due to the development, expansion and pervasiveness of pharmaceuticals, the establishment of healthcare systems and the ever-growing presence of medicalization in modern-day societies across the Western hemisphere.

Coinage of words and their eventual circulation across the spaces in which they are used: doctor's office, hospital, insurance company, pharmacy, patient's home, and school in the case of young patients, are not the end of the logistic flow of language.[29] In some cases, quite the contrary, they are just the beginning. Eventually, the logistics behind the creation of biomedical language, as is the case with any other logistic operation, is the final stage of a chain, beyond which there is little control. As I intend to show, there is a point at which terms such as "insulín" (*Ic.*) / "insulina" (*Sp.*) are appropriated and adopted socio-linguistically. It is not at a lexicological matter. Morphologically, the grammatical gender of Viagra, for example, the popular pharmaceutical used for erectile dysfunction, appears to be aleatorily and arbitrarily used by Spanish speakers, sometimes feminine and sometimes masculine. The lack of fixity in spoken language about the grammatical gender of pharmaceuticals is also identified in the case of above-mentioned Lantus. The diabetes-management medication is sometimes referred to as a feminine noun, "la insulina Lantus" or masculine "el Lantus".

Above all, what the examples have in common is that they show the sociolinguistic presence of biomedical and pharmaceutical concepts and trade names in everyday speech and how these have responded to historical processes involving science, commerce, and societies at large. Language thingifies pharmacy and pharmacy thingifies language, in both cases, so important information can be disseminated by producers, but also articulated and expressed by users. As observed by Halliday and Butt (2019), "[d]ifferent phases in human learning (common-sense – educational – technical – theoretical) all make their special demands on reasoning and persuasion in language" (p. 623). This entirely applies to the concatenation and embeddedness of pharmacy-related concepts, terms, and trade names. They go through stages that methodically can be traced from the laboratory to the shelves in hospitals, drugstores, pharmacies, supermarkets (in the case of over-the-counter products) and people's medicine cabinets. These words have a history that precedes the sociolinguistic dynamic in which they move. Knowledge about health is managed through the applicability of language to label referents to biomedical and therapeutic objects, but also through the management of a specific type of language that encapsulates it. Pun intended.

Returning to "Lantus" and "aspirin", these medicines ceased long ago being exclusive patrimony of science. As probably marketing teams of pharmaceuticals intended from the beginning, these are living words embedded in the living speech of speakers in an organic way, as much as in the organisms of patients who take them. As long as speakers of languages need medicines, pharmacy-related words will remain an integral part of their sociolinguistic reality. In one way or another, this will probably be the case for many more years to come. The plasticity of language at a structural level and of linguistic communities to adopt, adapt and appropriate changes seems infinite, but also somehow unpredictable. The case of Icelandic in this context is quite special. Iceland is a highly developed nation with a unique language that has managed to be preserved for centuries, partially because of the geographical location of the country. Eiríkur Rögnvaldsson (2022) reminds us though that the 21st century introduces a series of unexpected challenges for Icelandic in several ways. Technology and science are driving forces for the country as such, but also possible sources of obstacles for the prevalence of Icelandic (ca. 400.000 speakers) as the preferred means of communication. It demands "umburðarlyndi" (tolerance), notes Rögnvaldsson, nonetheless to realize that there are already many *types* of Icelandic. One of them is possibly the pharmacy-related parlance used by pharmacists, perhaps the vernacular used by patients who recurrently discuss the same medicine time after time or some of the authors of the following examples from the MÍM Corpus (2022):

17 Var á parkodín forte, voltaren og íbufen 400mg þegar ég bringubrotnaði fyrir einhverjum mánuðum og það gerði ekki skít (2007).
[*I was on Parkodín Forte, Voltaren, and Íbúfen 400 mg when I broke my chest a few months ago, and it didn't do a thing*]

18 Parkodin forte og íbufen rapid eru ekki endilega að virka svo vel ... (2013).
[*Parkodin forte and ibuprofen rapid are not necessarily working so well ...*]

In addition to the inconsistency in spelling, (i) [ɪ/ɪː] as opposed to (í) [i/iː], which are phonologically two similar and yet different sounds, in both 17 and 18, the interlocutors make a distinction between the normal brand and two enhanced versions ("forte" and "rapid" respectively). Based on the conversation, it appears as if these are brands well known to them.[30]

19 Ég lifi á íbufen 600mg og paratabs skv læknisráði þegar að mínir birtast ss verkirnir .... þá meika ég svona einhvernvegin daginn (2013).
[*I live on Íbúfen 600 mg and Paratabs as prescribed by my doctor when mine appear, i.e., the pain... then I somehow manage to get through the day*]

The patient in 19 alludes to the fact that ibuprofen constitutes part of his/her daily routine. The use of the possessive pronoun seems to emphasize the commensurability of experiencing pain often. Adapted word order (syntax) can be used to express scale.[31] In fact the opening sentence is indicative of how much relevance the medicine has in the life of the patient. Through exchanges of advice and personal experience, ibuprofen is described as a part of practices embedded in health-restoring ritualizations. As a means of categorization, adjectives about strength are used in combination with the adjectivized forms pharmaceutical companies use to brand enhanced or fortified formulas of medicines well known to users (e.g., forte). In the context of pain management, swiftness is of vital importance and hence, efficacy is contextualized in terms of pharmacokinetic efficiency (e.g., rapid):

20 Ertu búin að prófa að vera á íbufen nokkra daga og fara í sund og nuddpott (2013).
[*Have you tried taking ibuprofen for a few days and going to the pool and hot tub?*]

21 Og ef svo eru einhver verkjalyf sem ég get notað til að lækka verkin á meðan, íbufen virkar ekki (2013).
[*And if so, are there any painkillers I can use to reduce the pain in the meantime? Ibuprofen doesn't work*]

22 Eitthvað sem parkodin forte, panodil, paratabs og mjög sterkar íbufen hjálpa ekki (2014).
[*Something that Parkodin Forte, Panodil, Paratabs, and very strong ibuprofen don't help with*]

23 Eina eiginlega sem virkar er heitt bað eða hitapoki og svo íbufen (2014).
[*The only thing that really works is a hot bath or a heating pad, and then ibuprofen*]

The imprint of pharmacy in language is probably best illustrated in the expression "to be on something" (*Ic.* að vera á einhverju – dative case), as it is used in 20. The transitiveness of an action that at the same time implies the body is in its entirety under the influence of a pharmacological preparation, as opposed to the medicine being *inside of* the body. Strength, swiftness, and potency are, for obvious reasons, frequently identified in the corpora, either through adjectival or adverbial constructions, comparisons, and categorizations, such as "verkjalyf" (painkillers) in 21. The taxonomy of medicines, as it is expressed linguistically, naturally does not always overlap with pharmacological classification orders, protocols, or canons observed or used by pharmacists, physicians, or health professionals. In fact, as the next example illustrates, extracted from the PRESEEA Oral Natural Language Corpus in Spanish (henceforth PRESEEA Oral Corpus), even when it is used by professionals, pharmacy is linguistically articulated by speakers in their own terms. In the words of the interviewed speaker (E) who works in the health industry (Table 3.1).

An important distinction she makes is the difference between what she calls "remedios" (remedies) and "fármacos" (pharmaceuticals). Within the first category, she identifies "remedios fáciles" (easy remedies) and "remedios eficaces" (effective remedies), without implying that easy and effective necessarily overlap. The reflection on the prevention of health problems revolves around the agency and responsibility, in this case of caregivers. Perhaps the tone of the discussion prompts her to stress on

Table 3.1 PRESEEA Oral Corpus excerpt on remedies and drugs

GRAN_M23_010.MP3 PRESEEA Corpus 2004 Granada, Spain Informant: Female, 47 years old	
I: entonces si les das / remedios fáciles / y remedios eficaces que no son fármacos / eso no lo quieren ellos // entonces pues tú le dice <alargamiento/> s / pues / una cosa tan simple como es un lavado nasal // que haría que el niño tuviera la nariz siempre limpia / y la mayoría de las veces no se le infectaran los mocos esos y no necesitara antibióticos ni nada // son muy pocas las madres que lo hacen / lo primero que te dicen es que eso es imposible que el niño <alargamiento/> tiene mucha fuerza y que no <alargamiento/> // y / entonces yo le digo <cita>bueno pues si el niño tiene mucha fuerza entre dos personas uno lo sujeta y el otro le mete <alargamiento/> el suero y si chilla pues que chille // pero eso es un beneficio para el niño </cita>// <observación complementaria/> ¿qué pasa? que si le dan el jarabe / el niño <alargamiento/> y si encima de todo si sabe a fresa pue <alargamiento/> s / todos tan contentos y el niño <cita> ¡qué rico! <cita> y si puede el niño coge el bote el bote y se lo <alargamiento/> traga // pero <alargamiento/> // no llegan a ser las madre<alargamiento/> s muchas veces conscientes // o no llegan a entender que eso es fármaco / y es química que tú le estás metiendo / para que su hígado trabaje y demás eso no no le dan <simultáneo>ella s<simultáneo> //	I: So if you give them / easy remedies / and effective remedies that aren't drugs / they don't want those // so then you say <elongation/> s / well / something as simple as a nasal rinse // that would keep the child's nose always clean / and most of the time those mucus buildups wouldn't get infected, and they wouldn't need antibiotics or anything // very few mothers actually do it / the first thing they tell you is that it's impossible because the child <elongation/> is very strong and that they won't <elongation/> // and / then I say to them <quote> well, if the child is very strong, two people can handle it—one holds them while the other uses <elongation/> the saline solution, and if they scream, let them scream // but that's beneficial for the child </quote> // <complementary observation/> What happens? If you give them the syrup / the child <elongation/> and if on top of that it tastes like strawberry, well <elongation/> s / everyone's happy and the child <quote> "Yummy!" <quote> and if the child can, they grab the bottle, the bottle, and <elongation/> they gulp it down // but <elongation/> // the mothers don't often <elongation/> become aware // or they don't realize that this is a drug / and it's chemicals that you're putting into them / making their liver work and so on, but they don't give <simultaneous>they <simultaneous> care about that //.
E: <simultáneo> importancia<simultáneo> mira ¿y qué te ha parecido lo del proyecto de ley sobre el tabaco?	E: <simultaneous> importance <simultaneous> look, and what did you think about the tobacco bill?
I: espérate ¿cuál es el proyecto de ley?	I: wait, what is the bill about?
E: lo último / que está prohibido fumar / hasta en los restaurantes	E: the latest one / that smoking is prohibited / even in restaurants

the nature of treatments customary in biomedicine and the less desirable implications thereof. "Fármaco" in Spanish has a rather pejorative connotation, often linked to the negative image of large pharmaceuticals, profit-driven business models, prohibitive prices, lack of access for low-income patients, reported cases of adversary effects and the general idea that, despite their health-restoring effects, the negative impact on the human body might outweigh any possible benefit. The link between "fármaco" and "química" (chemicals) (*lit.* chemistry) underlines furthermore the distinction she makes at the beginning between "remedios" y "fármacos". Ontologically, they are not the same for the speaker and her language choice reflects this distinction.

The example also reflects the speaker's critique of the marketing techniques used to promote pharmaceuticals. While these methods are sometimes designed to aid patients, they can provoke skepticism. A "jarabe" (syrup) with an added strawberry taste, as she observes, risks seducing the palate of the young patient to consume more of it. As she explains, there needs to be awareness about the implications for the kidney to process "química" (chemicals) consumed in the name of health. Her observations about the enhanced flavor of the syrup resonate with the comments by Howard Ansel (1965) about the pursuit of originality that overrules the marketing of pharmaceuticals, only from a sensorial rather than linguistic vantage point. The observation about the use of flavor enhancers to make medicines more palatable is not very different from devising catchy brand names to make a pharmaceutical more memorable and popular.

The fact remains that, in the pharmacological sense, drugs are chemicals obtained from animals or plants, synthetized substances and in the case of biopharmaceuticals, engineered macromolecular products. A drug is a "chemical substance of known structure" that once is "administered to a living organism, produces biological effect" (Rang *et al.*, 2012, p. 1). Besides other ingredients (i.e., solvents), medicines are preparations made from drugs, often more than one. Despite the strict regulations, a high sense of mistrust towards pharmacological preparations is identified in many of the examples extracted from oral and written natural language corpora. As noted by Johnson (2016) in the context of Viagra, "it is not only the drug that prescribes behaviours or identities, it is decision makers, commercial actors, and medical experts who attach the drug to specific demands, images, and expectations to influence the behaviours of groups they are trying to govern, cajole or cure" (Johnson, 2016, p. 106). Due to the biosocial implications of preparations used for therapeutic purposes, they become cultural artifacts, nonetheless with linguistic implications. As I show in my discussion about Viagra in Chapter 5, sexual pharmacy does not only revolve around mitigating or resolving erectile dysfunction, but also around the words we use to refer to the pharmaceutical and how the object and accompanying words move across our conversations. Health-restoring and health-preserving objects, whether they are derived from pharmacy or from traditional knowledge (or any other type of medicine) encapsulate technoscientific innovation, epistemologies, but also entry points into the world of language as a method, system and means of communication. One, I suggest, that is logistically managed, planned and executed.

Notes

1 See Dyck and Charles (2021).
2 See Dauncey (2019) and Bennett and Balick (2014).
3 The name is derived from the alleged episode in which the Countess of Chinchón, the wife of Peru's Viceroy used the bark to cure her husband. A story that inspired Carl Linnaeus, who "wished to call it after the countess, but made a slight mistake in the spelling, naming the genus Cinchona, whence also the names cinchonine and cinchoninic acid" (Giral, 1966, p. 253).
4 As Gänger (2020) notes, "Spanish physicians and French housewives, Angolan surgeons and Portuguese merchants", on both sides of the Atlantic World, rendered cinchona "in their own language and refashioned" the history of the bark, ensconcing it "in their culinary habits, cultural imaginary or medical locality" (p. 182).
5 Documented as "Quino-Quino" in page 15 of *Quinología* by Hipolito Ruiz López (1792). A form less common than quina, but found for instance, "un bon moyen de guérir la blenuorrhagie, le mélange de baume de copabu et de quinquina, connu sous le nom de quino-baume" [*a good means of curing blennorrhagia, the mixture of balm of copabu and cinchona, known as quino-balm*] (Briquet, 1853, p. 211).
6 One of the most common is *Cinchona officinalis*, also known in English as cinchona bark, countess powder, Jesuit's bark, Jesuit's powder and Peruvian bark. In Spanish, besides Quechua-derived quina, other names include cargua cargua, cascarilla and corteza coja. In fact, "it was disseminated under the names of 'Countess's powder' and 'Jesuits' powder', because that order played a large part in its commercial distribution for medical purposes" (Giral, 1966, p. 252).
7 With major breakthroughs notably in France, Germany, and the United States, starting with the isolation of quinine by Pierre Joseph Pelletier and Joseph Caventou in 1820, followed by Paul Rabe's work on its synthesis in 1918 and its complete synthetization by Robert Woodward in 1944.
8 For the emphatic function of reduplication in Quechua, see Hoggarth (2004). See Hannß (2014) for a discussion on how the peculiarities of Kallawaya (secret language of healers) has possibly preserved aspects of Quechua reduplication from the past. For more on reduplication in Andean languages, see Hannß and Muysken (2014).
For repetition in poetry and prosodic contexts, see Mannheim (1998).

9 One could argue that pharmaceuticals have also their own identity in the sense that they are perceived for their traits (medicinal properties), uniqueness (two different formulas are different) and what they mean in the life of patients. In Johnson's words, "(w)hen an analysis of pharmaceuticalization teases out the actors behind a drug, one can see that a drug is much more than the pill that is claimed to cure a disease or alleviate a symptom" (2016, p. 107). Given the presence and impact of pharmacy in daily life, it seems obvious that medicines would have a special place in the therapeutic imaginaries, and by default, as I show in this book, this would have sociolinguistic implications at various levels.

10 This is the case in both Spanish and Icelandic. For a discussion on the historical adoption of "cocaine", among other Amerindian loanwords, see Erlendsdóttir (2015, 2017).

11 See Tsuda and Sano (1996).

12 Curare's long history resembles that of quinine, those looking for it "had the extract but no plant and later the plants but no pure compound" (Lee, 2005, p. 83).

13 On the specifics of the etymology of curare, entry in the *Etymology Dictionary* (2023) reads: "formerly also curari, curara, resinous plant substance used by South American natives for poisoning their arrows, later used medicinally as a muscle relaxant, 1777, from Portuguese or Spanish curare, a corruption of the name in the Carib language of the Macusi Indians of Guyana, wurali or wurari, which had a sort of click sound at the beginning, and is said to mean 'he to whom it comes falls' (when introduced into the blood the effect is almost instantly fatal)".

14 Dictionary of Mexican Spanish includes the verb "curarizar" to refer to "curarizar",v tr (Se conjuga como amar) (Med) Bloquear la transmisión de la unión neuromuscular por medio del curare o sustancias semejantes para reducir las convulsiones y espasmos musculares, o para obtener una más completa relajación de los músculos en la anestesia: Curarizar conejos despiertos [*v tr (Conjugates as amar) (Med) Block transmission at the neuromuscular junction by means of curare or similar substances to reduce convulsions and muscle spasms, or to obtain more complete relaxation of the muscles in anaesthesia: Curarize awake* rabbits] (online). The absence of "curarizar" from the RAE Dictionary recalls the repercussions of prescriptivism in the creation of dictionaries discussed by Del Valle (2009).

15 See Löfgren (1948).
Doctoral Disertation. Stockholm College. Stockholm: Hæggströms boktr.

16 Comparatively, "Natural cocaine is indeed a potent anesthetic but it also has powerful toxic effects; more especially, its stimulating influence on the central nervous system is extremely serious and may lead to subsequent addiction" (Giral, 1966, p. 237).

17 The origin of xylidine can be traced to Cahours (1850) who first coined the term: "Je le désignerai sous le nomme xylène [...]Le xylène mononitrique, étant dissous dans l'alcool et traité par le sulfhydrate d'ammoniaque, donne

une base analogue à la toluidine, que je désigne sous le nom de xylidine" [*I will designate it under the name xylene (...) Mononitric xylene, being dissolved in alcohol and treated with ammonia sulphide, gives a base analogous to toluidine, which I designate under the name of xylidine*] (p. 321).

18 About the reaction of the scientific community at the time, Hanson and Hingson (1950) noted, "Purely on the basis of efficiency, with a minimum of demonstrable toxic effects, xylocaine, introduced by Nils Lofgren' of Stockholm and first tested clinically by Torsten Gordh, anesthesiologist of Karolinska Stockholm hospitals, has commanded our respect" (p. 138).

19 Sanchez-Ramos (2020).

20 Translation taken from Barbara Johnson's annotated English version of *Dissémination* (1972).

21 See Frake (2012).

22 The type of dynamics I describe as bioclusivity, "the worldview held by the members of a specific group about nature, and which dwells in the interstices of what makes each culture unique" (Gonzalez Rodriguez, 2022a, p. xvi).

23 See Curzan (2014) on prescriptivism.

24 According to the *Guidance on the Use of International Nonproprietary Names (INNs) for Pharmaceutical Substances*, "-ol for alcohols and phenols" (WHO, 2017, p. 35).

25 Please note that aspirin is not capitalized because the settings of the dictionary embedded in the computer dictate that it is a common noun. The same settings demand though capitalization for both Lantus and Viagra. They are not (yet) part of the lowercase spelling repertoire.

26 A corpus of 7.6 billion words of data extracted from online newspapers and magazines published in 21 Spanish-speaking countries between 2012 and 2019.

27 It refers to the islets of Langerhans as main site of production that Sharpey-Schafer hypothesized to control the metabolism of glucose in the body. There is a controversy about the origin of the term. Jean de Meyer is credited for having coined "insuline" six years before Sharpey-Schafer. For more on the history on insulin, see Borgoño and Zinman (2012), Shampo and Kyle (2006), Sieck (2001), MacNalty (1964), Leikind (1948), Chattopadhyay and Sahana (2023).

28 For additional details on this adoption of medical insuline, see Vecchio *et al.* (2018), Quianzon and Cheikh (2012), Rosenfeld (2002), Bliss (1993), MacLeod (1978), Feasby (1958), Palmer (1946) and Banting (1929).

29 Here is important to note that my approach does not include medicines used in veterinary settings for two reasons. On one hand, because even if animals do show signs of communication, there is no evidence of them actively participating, from a linguistic viewpoint, in meaningful exchanges about name of medicines or pharmaceutical terminology. On the other hand, the analysis in this book is not post-humanist, as it focuses on processes intrinsically related to the type of agency exerted within and by human social structures and practices (i.e., collecting a prescription and buying a medicine). For more on the use of language in veterinary settings, see Mondémé (2011), Rando (2009), Sanders

(2007), Roberts (2004) and Sanders (1995). Animal language has been studied from various perspectives, see Cornips (2024), Meijer (2019), Radick (2007), Stibbe (2001), Hauser *et al.* (2002), Premack (1985), Sebeok (1965), Hockett (1959). As for plants, although there are important scholarly works about communication between plants, and with their surroundings, however there is a lacuna of human knowledge and understanding about linguistic strategies, patterns, or codes in the vegetal world.

30 Some studies have signalled an alarming use of certain medications in Iceland. At the turn of the century, there was already evidence of an "increase in the use of the prescription-only combinations containing 30 mg of codeine per tablet (Parkódín forte® and Panocod®)" (Almadóttir and Grímsson, 2000, p. 272).

31 See Sadegh-Zadeh, (2015).

4 Novelty, Lexicon and Medicines

Central to this chapter is an examination of the historical processes by which scientific discoveries and commercial outputs of pharmaceutical innovations were mediated through language. Specific focus lies, on one hand, on the adoption of knowledge about vitamins and their impact on human health in early 20th-century Iceland. On the other hand, the second part of the chapter reviews the commercial material prepared by European laboratories and intended to promote medicines among Spanish-speaking physicians in the 1950s, 60s and 70s. Both study cases illustrate the four-pronged context in which the language of pharmacy operates. Advances in the state of knowledge (science) materialize in the operationalized artifacts (medicines) that patients consume (commerce), but that they also need to understand or at least, be able to call by name (linguistics). As in the case of vitamins, the importance of pharmaceuticals for human health materializes in the dissemination of information designed to reach as many people as possible. Nonetheless because medicines are embedded in profit-driven constellations and networks informed by offer and demand. Effectively, commerce set off the vitamin "mania" era, as Apple (1996) explains, as eagerly as appetite for many other innovations derived from advances in pharmacy and promoted through an array of strategies at a global scale. One of them was to enhance the circulation of innovative, yet informative and convincing, leaflets that could be distributed among doctors. For mid-20th century European labs, it was relevant to make Latin American health professionals aware of new medicines and all their benefits. As with vitamins in Icelandic, the brochures themselves became nodes of dissemination of medical terminology in Spanish.

4.1 In Terms of Science

Vitamins in many ways revolutionized our understanding about the link between diet and disease. Knowing about their existence and understanding their role in living organisms logically resulted in the appearance of specific

DOI: 10.4324/9781003601654-5

terminologies. Names and concepts that, in the case of Icelandic, needed to be coined and adapted to the local linguistic protocols. It was not only a matter of expanding the Icelandic scientific lexicon. Alerting the nation about vitamins required the inception of vocabularies to contextualize publicly the ways in which nutrition played a role in the human body. Before vitamins became part of the pharmaceutical landscape, there had to be ways of talking about them. By reviewing the introduction of new terms in the aftermath of the discovery of vitamins, this section of the chapter offers some insights about the role of neologisms and translation in science communication and sociolinguistics of pharmacy-related terms.

In the 21st century, in most countries, nutrition-related terminology is easily found on the packaging of food, in promotional material about edibles, and as part of science communication campaigns endorsed by governments, national health directorates, and the private sector.[1] Information about the vitaminic content of a product has become a norm since the last decade of the 20th century. Providing consumers with details about the organic molecules present in their food is important. After all, these are essential micronutrients that the human body needs to function properly. Even if the quantities needed can be modest, obtaining vitamins from our daily diet is a precondition for staying healthy. Intake of vitamin A prevents, for instance, among others, night blindness, and keratomalacia, vitamin B_6 can fence off anemia and vitamin C, as history has taught us, is crucial to not develop scurvy. Due to their place in the nutritional label, vitamins are an integral part of the landscape (and health-sphere) of food. Today, by proxy, vitamins are also part of the linguistic landscape of shelves and one of the terms frequently cited to validate the dietic value of, for example, bottles of orange juice, breakfast cereal, dairy products, coughing pills and chewing gum. The popularity of vitamins probably explains also their presence in the promotional material of cosmetics, toiletries, and vaping e-cigarettes. Fortified consumables have a special place in the therapeutic and pharmacological imagination of consumers.

Vitamin (*Ic.* vítamín, *Sp.* vitamina) derives from "vitamine", a term originally coined by Polish-American biochemist Kazimierz Funk, whose name is commonly anglicized as Casimir Funk (1884–1967) in 1912. Funk sought to propose a term, based on his own understanding of the chemical composition of these molecules. He assumed that these were "amines", essential for life (*L.* vita) and hence "vit-a-mines".[2] But as Hopkins mentioned in his Nobel prize in Physiology or Medicine acceptance speech, the term was "agreed" on, hinting somehow at the fact that it did not suitably or conceptually reflect the nature of the substances to which it referred. Less than a decade later, Jack Drummond (1891–1952) suggested dropping the -e, as from "the strict standpoint of chemistry, not all the "vitamines" were chemical amines" (Semba, 2012, p. 77). The term differed considerably from "accessory factors (as I had then come to call them)" (Hopkins, 1929). Drummond's proposed system and the nomenclature of letters is still in use in the 21st century. The term was not adopted without resistance.[3] As Funk noted during a meeting held on 20 April 1916 at the New York Academy of Medicine:

> Besides the ordinary food constituents, such as proteins, fats, carbohydrates, lipoids, and inorganic salts, to which the value of our food is due, a number of substances can be found in small quantities which are as indispensable to life as the former constituents. These substances I have designated "vitamins," and the diseases which arise from their lack as deficiency diseases as "avitaminoses." I am surprised that the introduction of this term has aroused animosity since it is quite customary to give a name to a substance the presence of which is merely assumed before its actual isolation, and such terms are usually accepted until further evidence disproves their presence.
>
> (Funk, 1916, p. 1654)[4]

Like the encompassing term "vitamin", the etymological and morphological structure of the scientific names of each vitamin (and vitamer) bear earmarks of the history behind their discovery. An example thereof is "ascorbic acid" (vitamin C) ("a-" *L.* anti + scorbic > "scorbuticus" *M.L.* scurvy).[5] Its name serves as reference to the stories of many nations struggling against scurvy and the lives lost before finding out how to cure and prevent the devastating disease.[6] The recognition that some societies have been alert about the importance of ascorbic acid in their diet, centuries before its scientific identification, is in parallel to the recognition of local knowledges and their ancient rootedness in time, tradition and above all geographical space. A fact best exemplified by the encounter between a group of Saint Lawrence Iroquois and Jacques Cartier in the winter 1535–1536. Noticing that half of Cartier's crew was weakened by the lethal disease, the Iroquois shared with the French crew the knowledge about the properties of a local plant used to prepare an antiscorbutic concoction.[7] Cartier's crew survived and recovered completely.[8] The myth of the Tree of Life was born—one of many. Records about this encounter would eventually inform European scientists in their efforts to find a cure against the devastating disease. It led many to infer that the cure could derive from a substance found in nature.[9] Commonly, James Lind (1716–1794), is credited for the discovery of the efficacy of consuming citrus foods in the prevention of scurvy.[10] In *Treatise on the Scurvy*, Lind systematically covers the state-of-the-art medical knowledge on the disease, along with its "nature, causes and care", as the subtitle of the three-part volume reads (Lind, 1753).[11] The experiments conducted by the Scottish surgeon on board the Salisbury ship on 12 patients in the summer of 1747 paid off. Now there was finally confirmed evidence about a remedy to prevent sailors from developing the fatal condition. In line with Lind's recommendations, the British Royal Navy adopted as protocol the inclusion of lemons and limes as part of their vessels' provisions.[12] Understanding the prevention of scurvy lead to a better realization of the role of diet in the functioning of the body.

Advances in the identification of carbohydrates, fats, and the structure of proteins facilitated the design of experiments that could unravel the mystery of scurvy and its causes at a molecular level. Nikolaj Ivanovich Lunin (1854–1937) from the University of Dorpat,[13] was one of the first scientists to suggest the existence of unidentified factors present in food and fundamental for human health.[14] In general terms, Frederick Gowland Hopkins (1861–1947), is regarded as the main figure in the discovery of vitamins, although as he mentioned in his Nobel Prize acceptance lecture: "These substances, following the suggestion of Casimir Funk, we have agreed to call vitamins. Who was the 'discoverer' of vitamins? This question has no clear answer" (Hopkins, 1929).[15] But one thing was certain, the discovery of these organic molecules at different moments were all major scientific breakthroughs.

Throughout the 1920s, as in many other countries, Iceland underwent a period in which vitamins received particular attention. The science behind nutrition started to gain recognition, nonetheless because of the awareness about the many diseases that a nourishing diet could prevent. By 1924, advertisements of fortified food items started appearing in Icelandic print media. Such as the first addition of a brand of margarine containing A and B vitamins:

> Enginn gerir hið þjóðkunna »**Smára**« smjörlíki nema
> H.f. "Smjörlíkisgerðin" í Reykjavík.
> Í því eru A og B fjörefni (vitaminer).
> [*No one does the nationally known "Smári" margarine except*
> *"Smjorlíkisgerðin" Ltd. in Reykjavík.*
> *It contains A and B vitamins (Da.vitamins)*]
>
> (1924)[16]

The message is short, but quite insightful from a linguistic and semiotic perspective. Probably to avoid any misconception, the company promoting the fortified type of margarine Smjörlíkisgerðin (Margarine Factory), uses the Icelandic neologism "fjörefni" in combination with the Nordic term "vitaminer" within brackets.[17] Smári's margarine is being sold at a time when the Icelandic public is becoming more aware of the importance of consuming products containing vitamins. The advertisement kept appearing throughout 1924 and from 1925 onwards while the topic seems to gain more and more public attention. In May 1925, an author with the initials G.H. published a short piece in *Ísafold* on the properties of sunlight and cod liver oil:

> Eins og kunnugt er, hafa menn fundið það fyrir ekki alllöngu, að í flestri fæðu eru einskonar undraefni, sem nefnd hafa verið bætiefni (vítamin). Svo lítið er af þeim, að ekki hefir tekist að búa þau til hrein og þekkist því ekki efnasamsetning þeirra. Þó lítið sje af þeim, þá má líkaminn ekki án þeirra vera, vex ekki og sýkist ef þau vanta. Það er álit á hinar margra, að eitt af bætiefnum þessum forði börnum frá beinkröm og finnist í þorskalýsi. [*As is well known, it was discovered not long ago that most foods contain a type of wonder substance, called nutrients (vitamins). These substances are present in such small amounts that it has not been possible to isolate them in pure form, and their chemical composition remains unknown. Despite their small quantity, the body cannot do without them; it will not grow and becomes susceptible to disease if they are lacking. Many believe that one of these nutrients prevents children from developing rickets and can be found in cod liver oil*]
>
> (1925)[18]

The health benefits of cod liver were known to the Vikings, ancestors of Icelanders, but it was only in the 1800s that the tradition of consuming it as medicine revived in response to the sanitary threat posed by rickets. Now that vitamins had been discovered, it was possible to explain better the reason why a dietary supplement obtained from fish had therapeutic properties. As Bjarni Sæmundsson observed in an article published in 1923, that indeed "er lifrin úr þorski og ýsu líklega hollasta viðbitið, sem við höfum, jafn auðug og hún er að vítamínum" [*the liver from cod and haddock is probably the healthiest snack we have, as rich as it is in vitamins*] (p. 169).[19] Throughout the 1920s, the discussion about the role of diet in health was prompting an awakening about the physiological implications of food. One notorious inconsistency though was the use of a term to describe those "wonder substances" (vitamins) recently discovered, occasionally, simply called "materials" (*efni*).[20]

At the same time the notion of vitamins was a topic of conversation or a catchy term to sell food items in newspapers, a concerned scholar, Björg Þorláksson (1874–1934) saw the importance of informing Icelandic readers on the relevance of nutrition.[21] In order to reach the readers,[22] Þorlákson wrote two single-authored monographs, *Mataræði og þjóðþrif* (Diet and Hygiene) (1930) and *Daglegar máltíðir* (Daily Meals) (1933),[23] among other things, on the importance of food. By the time the books are published, there seem to be three competing, overlapping, synonymous and somehow interchangeable terms for vitamins in Icelandic: "fjörefni", "bætiefni" and "vítamín"/ "vitamin". The last two appeared in various combinations, mixing phonetically distinct (i) [ɪ/ɪː] and (í) [i/iː]. In *Mataræði og þjóðþrif*,[24] Þorláksson introduces a fourth possibility in her discussion about vitamin A, " 'vitamin'-efni" (1930, p. 194), although only superficially. As she clarifies in the beginning of the book, there is a reason to favor "fjörefni" over other possibilities: "Flestar húsmæður munu nú hafa heyrt talað um fjörefni, eða jafnvel og um leið hið útlenda alþjóðaheiti efna þessara:„Vitamin" [*Most housewives will have heard of fjörefni by now, or even at the same time the foreign international name for these substances: "Vitamin"*] (p. 30). Three years later, in *Daglegar máltíðir*, Þorláksson provides a full explanation for the preferred neologism[25]:

> Efni þau, sem ég nefni fjörefni á okkar máli, eru nefnd "vitamin" á alþjóðamálum, og nota margir læknar vorir þetta útlenda nafn, er þeir tala um fjörefnin — og jafnvel í riti líka. Álit ég þetta þó óhyggilegt, því ekki er hægt að ætlast til þess, að alþjóð manna *skilji* hið útlenda orð. Og engum verður það arðbært til þekkingarauka né hugljúft umhugsunarefni, sem hann eigi kann skil á, á þann hátt, að það geti vakið honum kunn hugsanasambönd og þannig leitt hann við hönd sér nokkur fet áfram á braut þekkingar og þróunar. [*The substances that I refer to as fjörefni (nutrients) in our language are called "vitamins" in international languages, and many of our doctors use this foreign name when talking about nutrients — even in writing. However, I consider this unwise, as one cannot expect the general public to understand the foreign word. And no one gains any benefit in knowledge or finds intellectual satisfaction in contemplating something they cannot comprehend in a way that might inspire familiar associations of thought and thus guide them a few steps further on the path of knowledge and development*]
>
> (1933, p. 5)

Þorláksson opposes the opacity of a word derived from Latin and scientific terminology, which she fears some will fail to comprehend. Since, the Latin word "*vita*, sem er aðalstofninn í orðinu vitamin, þýðir líf. Mætti því eins vel kalla efni þessi lífefni á okkar máli, ef aðeins væri hugsað um að þýða hið útlenda orð [*Vita, which is the root of the word 'vitamin,' means life. Thus, these substances could just as well be called 'life substances' in our language if the goal were simply to translate the foreign word*] (p. 5). However, there was a risk in adopting an opaque neologism, and "þá væri þó um of takmarkað innihald orðsins, og ófullnægjandi skýringin á *eðli* efna þeirra, sem um er að ræða" [*however, this would be an overly limited meaning of the word and an inadequate explanation of the nature of the substances in question*] (p. 5).[26] And since, even if vitamins provide energy, they do not do so "á svo háu stigi, að þau gefi dauðum líf—en þann skilning *mætti* leggja í orðið lífefni" [*to such a high degree that they give life to the lifeless — which is the interpretation one might assign to the word "life substances"*](p. 5). As mentioned before, "efni" (material) was a term used by many authors in print media at the time in their attempts to describe what vitamins were. The explanation continues with a remark about the fact that the word "fjör" also denotes "energy". An energetic horse, Þorláksson notes, is called "fjörhestur", and "fátt þykir góðum hestamönnum skemmtilegra en fjörugur hestur" [*few things delight good horsemen more than a spirited horse*] (p. 6). Conclusively, "þessi viðtæka merking orðsins fjör gerir það að verkum, að rétt er að þýða orðið vitamin með íslenzka orðinu *fjörefni*—eða og *fjörvi*, sem dr. Dungal notar" [*this broad meaning of the word "fjör" makes it appropriate to translate the word 'vitamin' into the Icelandic term fjörefni — or even fjörvi, as used by Dr. Dungal*] (p. 6).

The word "fjörvi" was not widely adopted in this context, but "fjörefni" certainly was. It is considered a synonym of "vítamín" in the Icelandic Terminology Dictionary (Íðorðabankinn). Before the publication of Þorláksson's books, the term had already appeared in several occasions in newspapers and magazines.[27] But few had gone in such a detail into the history of vitamins, their importance for human health and the applications of these discoveries in people's daily lives in Iceland before.[28] Other neologisms Þorláksson introduces in her books include "vöðvalopi" and "starfsefni", a translation of "vitasterin" and "vitasterol".[29] Whereas "vöðvalopi" still appears as an entry in historical dictionaries, "starfsefni" was never adopted as part of the lexicon surrounding vitamins.[30]

Based on the historical linguistic corpus, since the time Þorláksson's monographs came out, "vítamín" became progressively the preferred term in scholarly books, encyclopedias, textbooks, and popular science publications, as the following examples illustrate:

1 Enginn kann betur en hann, þegar allt kemur til alls, að meta þau vítamín, sem felast í sannleiksgeislum háleitra hugsjóna. *Skilningstré góðs og ills* (1939).
[*no one knows better than he, when all is said and done, how to appreciate the vitamins found in the rays of truth from lofty ideals*]

2 Eins og maður skyldi halda þau verkuðu þessi vítamín, sem þeir eru að gala um í Útvarpinu. *Blítt lætur veröldin* (1943).
[*Just as one would think those vitamins they keep raving about on the radio would work*]

3 Auk hvítu, fitu, kolvetna, vatns og steinefna, þarfnast maðurinn enn annarra matarefna,—vítamína, sem svo eru nefnd. *Bókin um manninn* (1946).
[In addition to proteins, fats, carbohydrates, water, and minerals, humans also require other nutrients — vitamins, as they are called]

The term "fjörvi" appears three times as a synonym, along with "bætiefni" (additive).[31] Noticeable examples of "fjörefni" over "vítamín" include: "Andrés skorti einhver þau **fjörefni** í sálina, sem skapa heilbrigt sjálfstraust" [*Andrés lacked some of the nutrients for the soul that build healthy self-confidence*] (1947) in which the term is used in figurative speech,[32] the encyclopedia *Alfræðasafn* and Nobel laureate Halldór Kiljan Laxness' *Heimsljós I-II* (1955), "Þetta nýa efni sem hann er kominn með, það heitir fjörefni og á að taka fram öllum efnum sem þekst hafa híngað til" [*This new substance he has discovered is called 'fjörefni' and is said to surpass all substances known up to now*](p. 30). As the first four examples illustrate, there is uncertainty about the recently adopted terms and their meaning and use:

4 Þegar þeir voru seztir að kaffidrykkjunni dró prestur flösku upp úr sál sinni og gaf þeim "fjörefni" út (1940).
[When they had settled down for coffee, the priest pulled a bottle from his bag and offered them "nutrients"]

5 Að í öllum hinum helztu fæðutegundum eru venjulega einkennileg efnasambönd, er skírð hafa verið vitamin, en á íslenzku bætiefni. *Matreiðslubók*. (1945).
[That in all the main types of food, there are typically distinctive compounds, which have been named vitamins but are called nutrients in Icelandic]

6 Vitamín (bætiefni) hafa uppleysandi og temprandi áhrif. *Grænmeti og ber*. (1945).
[Vita*mins (supplements) have a dissolving and tempering effect*]

7 Hefur orð þetta [::vítamín] verið lagt út á íslenzku á ýmsa vegu s.s. lífefni, bætiefni, þrifaefni, fjörefni og fjörvi, en samt er orðið vítamín alltaf notað jöfnum höndum (1946).
[This word [::vitamin] has been interpreted in Icelandic in various ways, such as lífefni (life substance), bætiefni (nutrient), þrifaefni (cleanliness substance), fjörefni (vital substance), and fjörvi (energy substance). Nevertheless, the word vítamín is still used interchangeably]

8 Kerti, ilmvötn, tannáburð, fjörefni, ungbarnafæðu (1946).
[Candles, perfumes, toothpaste, nutrients, baby food]

By 1951, there still seems to be disagreement on what the right term for vitamins in Icelandic should be. On the subject, Sigurður Pétursson observes that "vítamín, fjörefni, bætiefni, eða hvað það nú er kallað" [*vitamins, nutrients, supplements, or whatever they are called*] appear to be "á hvers manns vörum, og það í tvennum skilningi. Af þeim er alltaf meira eða minna í fæðunni, sem við neytum, og þannig hefur það verið frá alda öðli" [*On everyone's lips, and in two senses. They are always present to some degree in the food we consume, and so it has been since ancient times*] (p. 166). The popularity of vitamins as a topic of conversation is such that Pétursson has the impression that they are discussed "fram og aftur, menn láta í ljós áhyggjur sínar út af skorti á þeim, og menn spyrja um magn þeirra í matvælunum, sem þeir kaupa" [*back and forth, people express their concerns about a lack of them, and they ask about their quantities in the foods they buy*] (p. 166). But what is that that we call vitamins, he asks. The question is followed by a description of vitamins being organic substances that are found in natural foods in small amounts, but that have powerful effects on the consumer. By 1951, all 13 basic vitamins had already been discovered and the Vitamin Era,[33] as it has been called, had reached its peak. Notwithstanding, as Pétursson notes, Icelanders were using overlapping terms when talked about vitamins, with a marked preference for "vítamín", for instance, in encyclopedias and textbooks:

9 þær [::endurnar] verða sko að fá sín vítamín eins og aðrir. *Delerium búbónis. Gamanleikur með söngvum í þremur þáttum* (1961).
[*they [::the ducks] absolutely need their vitamins like everyone else*]

10 Vítamín í hrísgrjónum er aðallega í yztu lögum kornanna og kímveggnum. *Alfræðasafn AB 1-21* (1965–1968).
[*Vitamins in rice are mainly in the outer layers of the grains and the germ wall*]

11 Hann lagði til, að öll þessi efni skyldu einu nafni kölluð vítamín, eftir vita, sem táknar líf á latínu, og amín. *Alfræðasafn AB 1-21* (1965–1968).
[*He proposed that all these substances should be called vitamins, after vita, which means life in Latin, and amine*]

12 Mesta afrek hans var einangrun á B 12 vítamíni, sem hefur eitt ógnun þráláts blóðleysis (anemia perniciosa). *Alfræðasafn AB 1-21* (1965–1968).
[*His greatest achievement was the isolation of vitamin B12, which is a key remedy for the threat of persistent anemia (anemia perniciosa)*]

13 K-vítamín er nauðsynlegt, til þess að blóð geti storknað. *Alfræðasafn AB 1-21* (1965–1968).
[*vitamin K is necessary for the blood to clot*]

14 Nitur: Það kemur fyrir í fjölmörgum samböndum í lífverum s.s. próteinum, kjarnsýrum, og vítamínum. *Almenn vistfræði* (1975).
[*Nitrogen: It occurs in numerous compounds in living organisms, such as proteins, nucleic acids, and vitamins*]

By the 1980s, "vítamín" had become part of the Icelandic lexicon in contexts beyond entries in dictionaries, names mentioned on food labels or references to health-boosting vitamin pills. In the horoscope section of the Icelandic newspaper *DV* published on 22 January 1983, the brief forecast for the next day for those under the sign of Libra features the following prediction:

> Það gæti gerst í dag aö þú fengir góðar og nokkuð óvæntar fjárhagsfréttir, og jafnvel verið heppinn í spilum. Fréttirnar munu verka sem vítamínsprauta á þig og gera þig léttan í lund og drífandi í vinnu jafnt sem á heimili [*Today, you might receive good and somewhat unexpected financial news, and perhaps even get lucky in games of chance. The news will act like a vitamin shot, lifting your spirits and energizing you both at work and at home*].
>
> (p. 32)

The message is brief, but is symbolic for a discussion about the evolution of the language of pharmacy. As a reminder of the marker left by the discovery of food factors in the early 20th century "vitamin injection/shot" (*Ic.* vítamínsprauta) had been adopted in figurative speech to denote "stimulus", "impulse" or "encouragement".[34] Earlier examples of the expression extracted from the Ritmálssafn Corpus include:

15 Markið verkaði sem "vítamínsprauta" á Skagamenn (*Alþýðublaðið* 1963).
[*The goal acted as a "vitamin shot" for the Skagamenn team*]

16 En þetta vitakast verkaði *sem vítamínsprauta* á hana (*Alþýðublaðið* 1964).
[*But this burst of wisdom acted as a vitamin shot for her*]

17 Og köllin virðast verka sem vítamínsprauta á Reiff (*Morgunblaðið* 1968).
[*And the calls seem to act as a vitamin injection for Reiff*]

18 Markið verkaði sem vítamínsprauta á íslenzka liðið (*Morgunblaðið* 1974).
[*The goal acted as a vitamin injection for the Icelandic team*]

19 Virkaði sú kvikmynd sem vítamínsprauta á hippakynslóðina (DV 1982). [*That movie truly worked as a vitamin injection for the hippie generation*]

The use of "vítamínsprauta" in oral language is registered in the following examples taken from interviews transcribed for newspapers:

20 "það yrði nánast *sem vítamínsprauta* fyrir íslenskan lagmetisiðnað" (1976).
[*It would be almost like a vitamin shot for the Icelandic canned food industry*]

21 "Það virðist sem þessi tillaga mín í ríkisstjórn í gær jsess efnis að leggja fram frumvarp er bannaði flugmönnum verkföll fram yfir árarnót, hafi virkað sem vítamínsprauta á báða deiluaðila" (1979).
[*It seems that my proposal in the government yesterday to submit a bill banning pilots from striking until after the New Year worked as a vitamin shot for both parties in the dispute*]

22 "söngur er nefnilega eins og vitamin fyrir sálina" (1979)
[*Singing is, in fact, like a vitamin for the soul*]

23 "Ég tel að Jón virki sem vítamínsprauta á liðið eins og reyndar hefur þegar komið fram" (1979).
[*I believe that Jón works as a vitamin injection for the team, as has already been said*]

24 "Mikil hvatning er það starfinu að fá þess miklu gjöf úr Mývatnssveit sem getið er um á næstu síðum og hlýtur að virka sem vítamínsprauta" (1983).
[*It is a great encouragement for the work to receive this generous gift from the Mývatnssveit, mentioned on the following pages, which is bound to act as a vitamin injection*]

25 "Gjöf sem þessi virkar *sem vítamínsprauta* og hvatning á alla okkar starfsemi" (1991).
[*A gift like this acts as a vitamin injection and motivation for all our activities*]

26 "Að fá tvo blaðamenn í heimsókn virkaði sem vítamínsprauta á nemendur og kennara" (1994).
[*Having two journalists visit worked as a vitamin injection for students and teachers*]

27 "Aðgerðirnar gætu virkað sem vítamínsprauta í aðdraganda sveitarstjórnarkosninganna" (2014).
[*The "measures could act as a vitamin injection in the lead-up to the municipal elections*]

A final example is taken from a vacancy announcement placed by the Reykjavik City Hospital in a local newspaper in 1990:

> Við vonumst til þess að þú virkir sem vítamínsprauta á okkur og hafir samband sem fyrst.
> [*We hope that you will act as a vitamin injection for us and get in touch as soon as possible*]
>
> (Dagblaðið Vísir – DV, 28 May 1990: 41)

The term "vítamín" is also found in the MÍM 2022 Corpus, obviously with the connotation of organic substances with nutritious value and with the denotation of effective incentive or external stimulating catalyst, identified in prose and in examples of quotes transcribed from oral interviews:

28 "Þetta gerist ekki af sjálfu sér en enginn leggur þessa fyrirhöfn á sig nema hann finni gleði í því að gera þetta og finni sig uppskera í einhverju. Þetta er vítamín fyrir sálina" (1988).
[*This doesn't happen on its own, but no one makes this effort unless they find joy in doing it and feel they are reaping some reward from it. This is a vitamin for the soul*]

29 "En ég get lofað því að þetta er algjört vítamín fyrir sálina" (2004).
[*But I can promise you that this is an absolute vitamin for the soul*]

30 "Hrós er eins og vítamín fyrir sálina" (2016).
[*Praise is like a vitamin for the soul*]

31 "Slíkt vikaði sem vítamín þegar verið væri að berjast við sáralitla fólksfjölgun" (2000).
[*Such things acted as a vitamin when battling against minimal population growth*]

By then, one word, vitamin, either in plural or in combination with a figure has become part of the dietetic repertoire of terms used in daily conversations about the science of nutrition and the role of food in health. For over a century, in more or less degree, discussion about vitamins have been held in radio programs, televised talk shows and social media posts.[35] Since their discovery and eventual synthetization, vitamins have had a special relation with mass media, more so than other essential nutrients humans need to extract from their diets.[36] A point often dismissed from their brief inclusion in labels and various media formats (i.e., brochures, posters, television ads, etc.), is that the majority of vitamins are actually clusters of related compounds (i.e., vitamers). Hence, vitamin B_3, colloquially known as niacin, is in fact the generic descriptor of a family composed of three forms: Niacinamide, Nicotinic acid, and Nicotinamide riboside. This is important to realize the technical complexity behind the classification of vitamins and their eventual clustering as one single type of nutrient.

Vitamins have a linguistic history grounded in the processes behind the discovery, and eventual synthesis, and classification of the 13 known vitamins. A series of events that underscores the ways in which scientific terminology, commercial interests, and media play a role in the adoption, adaptation, and appropriation of medical language. From a pragmatic viewpoint, this holds relevance for specialized vocabulary, as well as language register. The use of terms, such as "ascorbic acid", or "riboflavin" as synonymous of, or in combination with "vitamin B_2", identified in the corpora, demonstrates how scientific language integrates into everyday conversation. The identification of "vitamin injection" (*Ic.* vítamínsprauta) in casual speech, used to refer to a stimulus or incentive in non-medical settings shows the plasticity of communities to respond to epistemic, scientific, and technological shifts.

In the Icelandic terminology dictionary (Íðorðasafn), a total of 17 entries are listed as terms derived from the lemma "vítamín". Linguistic evolution and vitality, best studied through samples of language in natural settings offers insights into the appropriation of science-driven lexicon. An example thereof is the semantic-pragmatic considerations behind the use of *vítamínsprauta* (*Ic.* vitamine injection)[37] to denote an incentive or a stimulus, e.g., in markets or in competitions. Despite Björg Þorláksson's admirable efforts, Icelanders are less prone to describe an incentive as "fjörefnasprauta".[38] What is certain is that the dissemination of information about vitamins, for instance given in the form of injections, is such that it has become a colloquial expression. As the next section shows, media outlets play a key role in creating networks to circulate pharmacy-related lexicon (i.e., epistemic labels).

4.2 Branding, Pharmaceuticalization, and Translation

The link between language and pharmaceuticals is illustrated at best in the strategies used to brand medicines. Even if the connection between health and pharmacology preceded notions of mercantilization and laissez-faire economics, the fact remains that pharmaceuticalization (overuse and increased use and availability of medicines) is profitable. Selling medicines is a multimillion-dollar industry and language is a tool in the toolkit of advertisement. Before patients can be prescribed medicines, doctors need to have basic information about the pharmaceuticals available in the market, such as their names, of course, but also their medical use and details about efficacy and pharmacology. Language plays a crucial role in medico-marketing, nonetheless because it relies on terminology that targets specialists. By reviewing a series of leaflets, the first part of this section explores how European laboratories linguistically paved the way for their products in Latin America and Spain during the second half of the 20th century. The focus lies on the role of commerce in the creation of linguistic strategies that could support their aspirations to conquer patients beyond the countries of production. As I show, the iconicity of the images was designed to compel, attract and seduce, but it was the scientific language that was meant to serve as decisive factor

The promotional material of Cibalgina for doctors, for instance, was one of several marketing strategies launched by Ciba (acronym of Chemische Industrie Basel), a Swiss chemical company, to advertise this medicine across Latin America in the 20th century.[39] The first records in Mexican print media date from 1927.[40] During the 1930s, it was promoted among physicians, for example, in *Gaceta Medica de Mexico* as one of the three Ciba products that "no deben faltar en el botiquín de urgencia del Médico" [*must not be missing from the doctor's emergency kit*] (1932, p. X), for being "el antidoloroso por excelencia" [*the pain reliever par excellence*], and more importantly, because "usando Cibalgina se ahorra en infinidad de casos la morfina" [*using Cibalgina saves the use of morphine in countless cases*] (1932, p. X).[41] As the leaflet produced in 1953 shows, two decades later, Cibalgina was promoted as a potent analgesic, effective for the treatment of "cefaleas, neuralgias, odontalgias, dismenorreas" [*headaches, neuralgia, toothache, dysmenorrhea*].[42] The "antidolores" (pain relievers) had turned into a "Analgésico poderoso de acción rápida y segura. Sedante y antipirético exente de alcaloides" [*Powerful analgesic with fast and reliable action. Sedative and antipyretic, free of alkaloids*] of which its margen terapéutico es muy amplio y no produce hábito" [*therapeutic margin is very broad and does not produce dependency*], available in "ampollas" [*ampoules*], "frascos de comprimidos" [*tablet bottles*] and "cajas de 5 supositorios" [*boxes of 5 suppositories*].[43]

Although the terms for headache (cefalea), neuralgy/neuralgia (neuralgia) and dysmenorrhea (dismenorrea) were technical, looking closely, "odontalgias" (toothache) particularly stands out, despite its long history. As a medical term, it was first used in the midwifery treatise *Diez previlegios para mugeres preñadas* written by Juan Alonso y de los Ruices de Fontecha in 1606 and who described it as "Odontalgia, el dolor de dientes" [Odontalgy, toothache] (p. 94).[44] According to Corpus del Diccionario histórico de la lengua española (CORDE Corpus henceforth), however, the term only appears in a written source again in 1807 and subsequently once in 1981 and twice in 1990. Its scarcity in the historical corpora, but not in medical treatises and textbooks, confirms "odontalgia" in Spanish has almost exclusively been used as part of a specialized and technical lexicon.[45] Coherently, the leaflet from 1953 includes the message "Literatura exclusive para médicos" [*literature exclusively for doctors*].

Assessing from the limited number of entries in the corpora, for an untrained reader, in any case, the terminology might not have been clear. From the perspective of its composition, the formula of Cibalgina underwent minor changes throughout the decades. The language used to promote the pharmaceutical kept changing as well, even if at its core, the medicinal use stayed roughly the same. The linguistic complexity of descriptors indicates a need for Cibalgina to retain its relevance. Also, a need for Ciba, its producer, to retain the visibility of its product within the repertoire of pharmacological options.[46] The words used in leaflets needed to *speak to* the intended readers and align with the lexicon used in medical treatises and textbooks. The language of pharmacy, as the example of Cibalgina shows, changes expeditiously, nonetheless because it revolves around by-products of science and commerce. In the process, language facilitates the process by which pharmaceuticals become referents in the therapeutic imaginary of a speech community, as the next examples from the CORDE Corpus show:

32 De las medicinas hasta que encontró unos supositorios de cibalgina bastante derretidos; se puso dos, porque la mitad se le quedó en los dedos, y volvió a la cama sujetándose las doloridas sienes con las manos (1988).
[*Among the medicines, he found some fairly melted Cibalgina suppositories; he used two, as half of them ended up on his fingers, and returned to bed, holding his aching temples with his hands*]

33 CIBALGINA, comprimidos, contra los dolores, fiebre, enfriamientos (5 comprimidos) (1993).
[*CIBALGINA, tablets, against pain, fever, colds (5 tablets)*]

34 Si usas Bisolvón, Cibalgina o Frenadol, puedes dar positivo por codeína (1994).
[*If you use Bisolvon, Cibalgina, or Frenadol, you may test positive for codeine*]

Whereas the second example is taken from a manual about camping practicalities (*Acampar. Manual práctico*),[47] the other two sentences appear in literary works. The first onc is a sentence from the novel *Amado Amo* by Rosa Montero and the third one comes from *El Alpe d'Huez* by Javier García Sánchez.[48] Based on the number of editions, for instance, of *Amado Amo*, by different publishing houses in Spanish, as well as in German, Romanian, Swedish, and Turkish,[49] the reception of the dark comedy has been positive since it was first published. *El Alpe d'Huez* has also been republished after its first release. Both García Sánchez and Montero inscribed Cibalgina as a notion in novels that might continue feeding the imaginary of many readers in present and future generations, even if the pharmaceutical disappears at some point from Latin American and Spanish pharmacopoeias. The literary references affix a certain sense of lexical permanence. As a brand, the name of the pharmaceutical expresses how it was perceived or bio-culturally experienced by the authors, who hoped it would convey a description and a descriptor to which readers would relate. Or, maybe a brand name that audiences would recall.

Someone like Teresa Claverol Mut, a former pharmacist who spent most of her working life as a pharmacist in Palma de Mallorca. In an interview, to a question about the most popular medicines requested by customers in the 1960s, she replies:

> Lo que más solicitaba era Optalidon en sobres, Cibalgina, Aspirinas, Buscapina compositum, y sobre todo hacíamos fórmulas magistrales, básicamente jarabes para la tos o fórmulas con quinina, ya que había mucha incidencia de fiebres palúdicas debido a los numerosos mosquitos de la zona de Sant Jordi, actualmente la zona del aeropuerto después de haber sido desecada. [*The most requested items were Optalidon sachets, Cibalgina, Aspirins, Buscapina Compositum, and above all, we prepared magistral formulas, mainly cough syrups or formulas with quinine, as there was a high incidence of malarial fevers due to the numerous mosquitoes in the Sant Jordi area, which is now the airport area after being drained*]
>
> (Muntaner i Mariano, 2008, p. 90)

Claverol Mut's memory does not seem to fail. According to quantitively obtained data, by the end of the 1970s, Cibalgina, along with Optalidón and Nolotil, produced respectively by Ciba, Sandoz, and Europharma, were the non-narcotic analgesics capturing 64% of the sales in the Spanish market (Lobo, 1977). In hindsight, compared to the 1970s and 1980s, Cibalgina might be less present pharmacologically in Latin America and Spain, but linguistically, it retains its status as a brand name and referent in Spanish commercial print media, encyclopedias, medical treatises, catalogs, and the above-mentioned novels. The name of an analgesic, not surprisingly, is likely to become a household brand and retain, in a certain way, its sociolinguistic identity. Particularly, if a comparison is drawn with discarded medicines or those briefly marketed or used for more specific purposes and in specific contexts, such as the case of Vioftalyl, discussed in the introduction. In contrast with an analgesic, as popular as Cibalgina, Vioftalyl has disappeared from the sociolinguistic and sociocultural repertoire. In cases where pharmaceutical technologies trigger biosocial, political and lifestyle revolutions, language, and pharmacological practices seem to merge.

Throughout the 1960s, contraceptives in one form or another were becoming part of the informal therapeutic repertoire in several Spanish-speaking countries and Western laboratories saw the potential of local markets for birth control and menopausal therapy pharmaceuticals.[50] One of them was Organon, a Dutch company, still today based in Oss, the Netherlands, and endorser of "Método 22" (Method 22). A prescription-based regimen of 22 pills containing ovulostatic agents is to be taken sequentially every day, followed by six fixed "rest days".[51] An approach that "llena en forma óptima la exigencia de combiner un efecto seguro con un método sencillo" [*optimally meets the requirement of combining a safe effect with a simple method*] (Organon, 1966), according to the leaflet distributed among physicians.

Lindiol 2.5 were tablets containing, in line with its name, 2,5 mg of linestrenol and 0,075 of mestranol. An easier, simpler, more practical, and more effective ovulostatic agent, as the brochure states. From a linguistic perspective, two parts of the text stand out: "ovulostatico genuino" (today "ovulostático") [*genuine ovulostatic drug*] and "efectividad ovulostatica infalible" [*infallible ovulostatic effectiveness*]. Above all, because as Rodríguez-Ocaña *et al.* (2012) observe in the case of Spain, pharmaceutical terminology creatively "eludía la definición como contraceptivo, con lo que no infringrían la ley española en vigor" [*eluded the definition as a contraceptive by which it would not infringe current Spanish law*] (p. 486). Labeling a pharmaceutical as a contraceptive was simply not allowed. The Mexican case was not different as the prohibition of the sale and advertising of birth control therapies remained banned until 1973 (Secretaría de Salud, 2014).[52] Medicines were advertised using terminology that would describe the pharmacology and chemistry of their preparations, not necessarily their typology or classification. Other terms commonly identified in scientific and medico-marketing literature of the time included "gestágenos" (gestagens) and "anovulatorios" (anovulation-inducing medicines) (Rodríguez-Ocaña *et al.*, 2012), both used as nouns and adjectives and as descriptors of types of birth control medicines.

As the promotional material of Lindiol 2.5. indicates, that in the context of contraceptive pharmacy, language is managed and strategically planned to circumvent possible legal repercussions. Birth control regimens were also socially stigmatized as they contradicted prescriptive moral canons endorsed by Catholicism. Language served to create a space in which medicines could circulate in parallel with the cultural dimension of confessional attitudes about contraceptives. For many of the Western laboratories distributing their innovations across the Spanish-speaking world, it was a matter of supplying a growing market. Terminology was part of the logistic operation behind the circulation of Lindiol 2.5. throughout Latin America.

A few years after Mexico, advertisement of contraceptive treatments eventually became legal in Spain in 1978 (Folguera, 2022). Throughout the 1970s, in other parts of Latin America, drastic changes had taken place in the biosocial climate in regard to birth control. As Felitti (2022) has shown, several social processes, involving various actors and agendas had an impact on the uneven adoption of contraceptive policies across the region. An ideological shift and several events shaped the conditions for the implementation of family planning programs. By the 1980s though, the "la píldora"/"la pastilla" (the pill), even if only as a possibility, had been incorporated in the Spanish-speaking pharmaceutical and cultural imaginary.[53] An example thereof is the following conversation registered by the PRESEEA Oral Corpus of an interview taking place in La Habana (Cuba) in 2012 (see Table 4.1).

Through the answer provided by the grandmother in the interview about her granddaughter's upbringing, it becomes clear that the notion of "la píldora" is engrained in the medical, sociocultural, and sociolinguistic repertoire of both women, despite their generational gap. In her words, the age difference prompts her to comment on her expectations of pharmacy as a means to mitigate what she considers to be a social issue. For the informant, pregnant 14- and 15-year-old young girls have no future. In line with this vision, the *future* lies in adopting pharmacological solutions for adolescent pregnancy, which has social implications at various levels. "La píldora" escapes any semantic, pragmatic, rhetoric, and discursive frame.[54] It concatenates pharmacy, society, and language in a colloquial label transversed by the biocultural implications of gender, religion, and body rights—but also—commerce. Worldwide, birth control pills are believed to be a multimillion-dollar industry. In 2015, 15% of the women in Latin America used the pill, as compared to nine percent of the women worldwide and "the pill" "accounts for at least ten percent of contraceptive practice in over 70 percent of the countries with sufficient data to enable estimates" (UN, 2015, p. 24).

The high numbers underline nevertheless the fact that "l'industrie pharmaceutique du début du siècle commença à se préoccuper davantage

Table 4.1 PRESEEA Oral Corpus excerpt on "the pill" as a contraceptive method

LHAB_M32_067.MP3 PRESEEA Corpus 2011 La Habana, Cuba Informant: Female, 57 years old	
E: <tiempo = "39:18"/> ¿y qué temas le preocupan con su nieta que es adolesce<alargamiento/>nte porque<alargamiento/> / está el tema de<alargamiento/> la mo<alargamiento/>da de los / la educación en los jó<alargamiento/>venes?	E: <time = "39:18"/> And what issues concern you regarding your granddaughter who is an adolesce<elongation/>nt because<elongation/> / there's the matter of<elongation/> the fa<elongation/>shion among / the education of yo<elongation/>ung people?
I: no tanto de la moda // no tanto de la moda porque / yo también fui joven / también tuve quince años y me puse minifalda / maxifalda / pantalones corte chevalier y corte campana // no es tanto eso // como / la poca <silencio/> no te voy a decir que / que cultura porque // Cuba es un país de gente culta // pero sin educación // porque la gente ahora no son educadas<alargamiento/> / en ninguno de los sentidos // la juventud inclusive eeh en mi tiempo por ejemplo // que no es un tiempo tan lejano // una muchachita en la calle decir una palabra / obscena como digo yo // y ahora le dicen como si fuera tomarme un vaso de agua / lo mismo hembras / que varones / no importa / ni siquiera importa el lugar donde estén // yo hace unos días pasé por frente a la escuela de mi nieta // de la secundaria donde ella estudia / y me tuve que parar y retroceder y decirle a una niña <cita> pero ven acá muchachita tú no te das cuenta que tú eres una hembra / que tú no eres un varón // si en los varones se ve feo // ¿qué quedará para ti? </cita> // y eso de / <cita> mi tía dame un cigarro // préstame un fósforo </cita> / ¿qué cosa es eso? / eso no es una cosa que tenga que hacer la juventud / la juventud está para estudiar // para	I: not so much about fashion // not so much about fashion because / I was young too / I was fifteen too, and I wore miniskirts / maxiskirts / chevalier-cut pants and bell-bottoms // it's not so much that // as / the lack of <silence/> I'm not going to say / culture because // Cuba is a country of cultured people // but without education // because people now aren't educated<lengthening/> / in any sense // even young people in my time, for example // which isn't so long ago // a girl in the street saying a word / obscene as I call it // and now they say it as if it's like drinking a glass of water / both girls / and boys / it doesn't matter / it doesn't even matter where they are // the other day I passed by my granddaughter's school // the secondary school where she studies / and I had to stop, turn back, and say to a girl <quote> but come here young lady don't you realize that you are a girl / that you're not a boy // if it looks bad in boys // how much worse does it look for you? </quote> // and this thing about / <quote> auntie, give me a cigarette // lend me a match </quote> / what is that? / that's not something that young people should do / young people are for studying // for having fun / but having fun in a very healthy way // right now they gather on the corners, and sometimes it's just

Table 4.1 (Continued)

divertirse / pero para divertirse muy sanamente // ahora mismo se ponen en las esquinas que hay veces que da <énfasis> pena </énfasis> / y vuelvo y le repito que yo fui joven // hice todas las cosas que hace la juventud // pero en una esquina que parece que lo único que le falta es una cama // no hay necesidad de eso <silencio/> por eso es que a los muchachos tienes que abrirle los ojos desde que empiezan // esto es por esto / y aquello por lo otro / y aquello por lo otro / porque // se habla de mucha educación sexual / pero hay que hablarlo con crudeza // no adornando la píldora como digo yo // decirle esto es por esto / por aquello / y por aquello // si después lo hacen // lo hacen porque quieren no porque no lo sepan <silencio/> y eso sí es una cosa preocupante // las cosas que están pasando ahora yo nunca había visto como ahora / en este tiempo // muchachitas / de catorce y quince años / embarazadas // ¿qué futuro tiene esa niña? // que si estaba estudiando no tiene ningu <palabra_cortada/> no ha podido terminar ninguna carrera // que a la vez que tienen un hijo // no tienes madurez y no tienes responsabilidad / es un niño / con un niño / para pasar trabajo // porque todas / no son todas las / las abuelas / que pueden asumir un muchacho // ¿y por qué pasa eso? / por la poca / comunicación // no de que te lo digan por televisió<alargamiento/>n ni te lo digan por radio / <énfasis> de la misma familia </énfasis> // que la misma familia tiene que / desde que los niños / tienen eeh / razonamiento por ejemplo // tienen que inculcarle miles de / miles de valores / miles de cosas que ya eso no se hace // para que<alargamiento/> no vayan a / a fracasar // y ahora eso es una cosa que es lo más normal de la vida / tú vas a un hospital	<emphasis> shameful </emphasis> / and I'll say again that I was young // I did all the things young people do // but on a corner that looks like the only thing it's missing is a bed // there's no need for that <silence/ > that's why with young people, you have to open their eyes from the start // this is for this / and that for that / and that for that / because // there's a lot of talk about sexual education / but you have to talk about it bluntly // not sugarcoating it as I say // tell them this is for this / for that / and for that // if they do it afterward // they do it because they want to, not because they don't know <silence/ > and that is something that's really worrying // the things happening now I've never seen before / in my time // girls / fourteen and fifteen years old / pregnant // what future does that girl have? // if she was studying, she doesn't have any<word_cut/ > she hasn't been able to finish any career // and when they have a child // they don't have maturity, and they don't have responsibility / it's a child / with a child / to struggle through life // because not all of them / not all of the / grandmothers / can take care of a child // and why does that happen? / because of the lack of / communication // not what they tell you on televisio<lengthening/>n or on the radio / <emphasis> from the family itself </emphasis> // the family itself has to / from when the kids / have eeh / reasoning, for example // they have to instill thousands of / thousands of values / thousands of things that don't get done anymore // so that<lengthening/> they don't end up / failing // and now it's just one of the most normal things in life / you go to a hospital // gynecological-obstetric / and you find / thousands of young girls / sometimes even in their uniforms to get a D&C

(*Continued*)

Table 4.1 (Continued)

// ginecostétrico / y te encuentras / miles de niñitas / hay veces que hasta con uniforme para hacerse un legrado <silencio/> <ruido = “carro”/> no vamos tampoco a echarle la culpa // a / a la situación que estamos viviendo no / vamos a echarle la culpa a la poca formación que le han dado a esa niña // que a lo mejor // la madre se ocupa // entre comillas / o no se ocupa // y como la chiquita no tiene esa gran<alargamiento/> / ese gran conocimiento / imagínate tú // ¿qué es lo que puede pasar? / yo hay veces que las miro y digo <cita> ¿y esto qué cosa es? </cita> // porque ves una niña con un niño // ¿qué <ininteligible/> formación podrá tener ese pobre chiquito? // si lo está criando / una niña / igual que él <silencio/> y esa es una de las cosas que es preocupante <ruido = “manos frotándose”/> // pero preocupante / preocupante // sobre todo que ya te estoy diciendo con las que están saliendo ahora como mi nieta que tiene trece años que si yo la dejo de la mano <ruido = “manos frotándose”/> je / ¡por favor! // y es también que se reúnen a / a conversar / y siempre no conversan cosas buenas // porque / para que el mundo sea mundo tiene que haber de todo / buenos y malos y regulares // pero si tú no empiezas a sembrar desde chiquito / después no puedes recoger //	<silence/> <noise = "car"/> and we’re not going to blame it all // on / the situation we’re living in, no / we’re going to blame it on the poor upbringing that girl had // maybe // the mother takes care // quote-unquote / or doesn’t take care // and since the little girl doesn’t have that great<lengthening/> / that great knowledge / imagine that // what can happen? / sometimes I look at them and I say <quote> and what is this? </quote> // because you see a girl with a child // what <unintelligible/> upbringing can that poor little child have? // if they’re being raised / by a girl / just like them <silence/> and that’s one of the things that’s concerning <noise = "hands rubbing"/> // but concerning / concerning // especially what I’m telling you now about the ones who are coming up, like my granddaughter, who is thirteen, and if I let her go unchecked <noise = "hands rubbing"/> jeez / please! // and it’s also that they gather to / to talk / and they’re not always talking about good things // because / for the world to be the world, it has to have everything / good and bad and average // but if you don’t start planting seeds when they’re little / then you can’t reap anything later //
E: <tiempo = “43:42”/> ¿y qué valores le ha inculcado a ella? //	E: <time = "43:42"/> And what values have you instilled in her? //

de problèmes auxquels elle pouvait proposer une solution” [*the pharmaceutical industry at the beginning of the century began to focus more on problems for which it could offer a solution*] (Giroux, 2007, p. 59). To the extent that “mauvaise haleine, les problèmes de silhouette, de fatigue et l’écoulement nasal sont devenus des problèmes de santé” [*bad breath, figure problems, fatigue and runny nose have become health problems*] (p. 59).

Language has played an important part in the creation of conditions and diseases with convincing names and simultaneously, the introduction of medicines with equally convincing names and supportive terminology. After all, "l'émergence d'un discours sur le risqué" [*the emergence of a discourse on risk*] (p. 59) stems from the ability to communicate and convince the public as well as physicians. Not surprisingly, those who show concerns about the so-called "disease mongering" suggest the need for a "linguistic de-escalation" (Pariente and Laharie, 2014), implying that language holds a key to creating equilibrium between the real needs of patients in terms of pharmaceuticals and the overprescription of medicines for the sake of increasing profits. Names, concepts and terms are crucial. Today, Cibalgina, Lindiol 2.5 and Vioftalyl might have been labeled differently as "inventing names for drugs has become a big business" (Gangwal and Gangwal, 2011, p. 3). Nowadays, branding agencies are involved in the crafting of "a powerful, unique name that will entice physicians and patient" that can phonetically increase the branding effect. Examples thereof include [b] to denote reliability and slowness, [f]-[v]-[z] to connote speed, being [z] particularly "powerful" as a marketing tool (4). Coining the name of a medicine step-by-step for it to reach as many users as possible reinforces the observation that the language of pharmacy operates as a linguistic operation. As Enfield and Sidnell (2022) assert, "when we use language, we are not merely identifying or drawing attention to referents, we are also thematizing and characterizing them" (p. 133). Pharmacy is articulated linguistically through names that convey and appeal to general perception about science, deemphasizing the monetization aspect of medicines. More importantly, outputs of pharmacy become part of the artifacts that move across society prompting in the process a reaction at legal, cultural, and linguistic levels. Language registers and allows speakers to adapt, adopt and articulate the input of medicines in their bodies, lives, and communities. As I suggest in the next chapter explores, pharmacy creates linguistic by-products.

Notes

1 Packaging of medicines, as many other consumables operate as multilingual hierarchical spaces. See Sebba (2023) for an insightful study on the linguistic implications of packaging in the context of commerce, information and ideologies.

2 Amines refer to "a class of nitrogen- containing organic compounds derived from ammonia [NH_3]"; the term provides the basis for the phrase "amino acid" (Semba, 2012, p. 77).

3 As mentioned in Chapter 1, several names preceded "vitamin", such as "trophone", "sitacoid", "vitaaids", "biogens", "advitants", "ergotropes", "ergogents" and "accessory food factors" (Ramsden, 1918; Brown, 1920; Maltz, 1992).

4 Statement during a meeting held on 20th of April 1916 at the New York Academy of Medicine; an event chaired by Dr. Walter B. James.

5 The etymology of "scurvy" in English is not entirely clear. There is a suggested connection to French scorbut and Dutch scheurbuik, probably related to the Old Norse term skyrbjugr (*Ic.* "skyrbjúgur"), explained by some as an agglutination of a bjugr (swelling) and skyr (sour milk): "swelling caused from drinking sour milk on long-distance voyages" (OED). Alternatively, its suggested meaning relates to Middle Low German, "buk" (belly), or Middle Dutch "buik" in combination with Middle Low German and Dutch "schoren" (lacerate): "the disease that causes lacerations in the belly". Spanish "escorbuto", according to most sources, can be traced to French "scorbut", presumably an adaptation of a Dutch loanword. Carpenter (2012) suggests that "scorby" preceded "scurvy" in English.

6 It was especially acute among those travelling long distance on sea. In travel literature from before scurvy was properly described and understood, the origin of the term remains vague. Pinkerton (1808), for instance, is not entirely sure that the oldest reference to "skyrbjugur" from 1289 in Icelandic and Norwegian sources entirely clarifies whether elephantiasis and scurvy are the same or different diseases.

7 This is considered as "One of the first documented uses of Indigenous medicine in North America" (Durzan, 2009, p. 1). The term antiscorbutic (i.e., "without scurvy") is attributed to a coinage by Johannes Bachstrom in 1734 (Price, 2017).

8 See Buchanan (2021) https://daily.jstor.org/plant-of-the-month-tree-of-life/.

9 His book includes a detailed review of all knowledge related to curing scurvy using external agents.

10 Even if there is evidence that Agustín Farfán, recommends in *Tratado Breve de Anatomía y Cirurgía, y de Algunas Enfermedaded*, published in 1579, the consumption of lemons and oranges to treat scurvy.

11 On the origin of the "scurvy", Lind (1757) observes that "the word seems to me most naturally to be made out from 'scorb' in the *Sclavonic* language, which signifies a disease; this being the endemic evil in Russia, and those northern countries, from whence we borrowed the *name*" (p. 289).

12 Although this was only done many years later, most likely by intervention of Gilbert Blane, Commissioner of the Sick and Wounded Board, who in a short book entitled *A short account of the most effectual means of preserving the health of seamen, particularly in the Royal Navy* (ca. 1780–1781), wrote: "It must likewise be observed that such is the excellence of those articles of food and medicine, particularly the malt, lemons, oranges, and sourkrout, that they are efficacious in curing as well as preventing the Scurvy†" (Blane, 1780-81, p. 54).

13 A German-speaking university that operated between 1802 and 1893 in the premises of current-day University of Tartu (Estonia), and originally founded as a Swedish educational institution under the name Academia Gustaviana.

14 Lunin's publication, "Ueber die Bedeutung der anorganischen Salze für die Ernährung des Thieres" (On the importance of inorganic salts for animal nutrition) (1881), provided some of the first clues about these unidentified

substances. The implications of the experiments carried out by Lunin, Gustav von Bunge, his mentor, and Carl Socin were not pursued any further by the authors of the experiments and published results (Semba, 2012).

15 Besides the other recipient of the Nobel prize in physiology-medicine that year, Christiaan Eijkman (1858–1930), Hopkins mentions in his speech the contributions made by Cornelius Pekelharing (1905), and above-mentioned Bunge, Lunin and Socin. Paul F. Ostrowski (1986), observes to the question, "Who discovered vitamines?", and after reviewing much evidence that, Eijkman's discovery of a deficiency disease in birds led to the discovery of thiamine, but it was Funk's work which first established that vitamines were a new class of essential nutrients, and stimulated research throughout the world on them. Casimir Funk is, therefore, the discoverer of vitamins (p. 183). On the term, he adds that since "vitamine" does not provide any additional information on the chemical nature of these substances, in his opinion, "Funk's original spelling of vitamine should be retained" (p. 181). Takaki Kanehiro's (1849–1920) work on beriberi in 1884 is often forgotten as an important early contribution to the history of vitamins. He correctly established that there was a link between human diet and developing the disease. Mistakenly, he assumed that the specific physiological cause was a deficiency of protein (Sugiyama and Seita, 2013).

16 The add first appeared on the 1 March 1924 in *Tímarit Verkfræðingafélags Íslands* (Magazine of the Icelandic Guild of Engineers).

17 Due to the historical connection, most likely Danish.

18 The article "Sólarljós og þorskalýsi" appeared in *Ísafold* on 14 May 1925.

19 Bjarni Sæmundsson was an authority in the topic of Icelandic salmon, author of *Íslensk dýr I. Fiskarnir* (*Icelandic animals I. Fish*), published in Reykjavik in 1926. Along with Knut Dahl, he carried out pisciculture-related research in the early 20th century.

20 See for instance the discussion in the article "Vitamin"-kenningin ("Vitamine" Theory), published in the periodical *19. Júní* in November 1924.

21 Björg Þórlaksson was the first Icelandic woman to complete a PhD degree. She graduated from Sorbonne University in Paris in 1926 (Einarsdóttir, 1984).

22 Specifically, as the author explains in the preface, "Bók þessi er hin fyrsta manneldisfrœði, sem rituð hefir verið fyrir íslenzkar konur" [*This book is the first work on nutrition written specifically for Icelandic women*] (Þorláksson, 1930, n.p.].

23 Þorláksson (1933) explains in the prologue that *Daglegar máltíðir* is written in the hope that "húsmæður geti notað sér þekkingu þá á nauðsyn og nytsemi fjörefna og þekkingu þá á samhæfni ýmsra flokka næringarefna og ýmsum höfuðþörfum líkamans, að ég ræðst í að láta þessi erindi, sem upphaflega voru flutt í útvarpaniu, koma á prent" [*Housewives may benefit from this knowledge about the necessity and usefulness of nutrients, as well as the understanding of the compatibility of various types of nutrients and the body's fundamental needs. For this reason, I have decided to publish these talks, which were originally delivered on the radio, in printed form*] (p. 4).

24 A book Þorláksson wrote in Paris in the winter of 1929.

25 Intellectual figures compelled to advance scientific fields and scholarly work in Iceland, as shown in Chapter 2, realize that the most practical solution to

fill the void in terminology before a body of knowledge can be adopted is for them to engage in the coinage of terms. As Ari Páll Kristinsson (2018) observes on lexicon in Icelandic, "few terminology committees, and some groups of translators and text book authors, are very active in the systematic lexical cultivation of particular fields of expertise, for example in medicine, physics, computer science, trade, education theory and biology (in particular terminology for plants and fish" (Kristinsson, 2018, p. 246).

26 Still today is the case that: "It is quite common for ordinary people to come up with ingenious ideas for new lexical items (purist neologisms), either for fun or out of necessity, e.g. to attain a particular goal" (Kristinsson, 2018, p. 246).

27 See for instance the articles "Heilbrigðistíðindi" in Morgunblaðið (29 September 1925), "Af Ísafirði" in *Morgunblaðið* (7 August 1926) and "Breyttir lífnaðarhættir" in *Lögberg* (26 May 1927). Interestingly, "Af Ísafirði" introduces the term "vitamina" as an alternative translation for "fjörefni".

28 Kristmundsdóttir (2001) asserts that Þorláksson's personal life and her position as woman heavily influenced her career as scholar negatively, even if her contributions as translator and lexicographer were many and large in scale, she was not always credited for her work (e.g., in her contributions to the 20-year-long Danish-Icelandic dictionary project). Kress (2002) observes that Þorláksson faced also obstacles trying to publish translations of literary works and received little attention and recognition despite the quality of her work.

29 Whereas "vitasterin" was the term used in German scientific circles, "vitasterol" was the common term in scholarly articles published in English- and French-speaking countries. See Hermann and Zentner (1933), Rabl (1931) and Heubner (1930) for a discussion about vitasterin at the time.

30 On the impact of Þorláksson's impact in the Icelandic intellectual community, Kristmundsdóttir (2011) asserts that "Miðað við ríkjandi hugmyndir um geðsjúkdóma eru slík veikindi til þess fallin að varpa rýrð á framlag fólks á hvaða sviði sem er. Hvort sem það var af þeim sökum eða öðrum sem fræði Bjargar og hún sjálf féllu í gleymsku skal ósagt látið. Fræðum Bjargar var í öllu falli sópað út af fræðasviðinu og höfðu því að öllum líkindum takmörkuð áhrif ef undan eru skilin rit hennar um næringarfræði ætluð íslenskum húsmæðrum. Kerfisbundin könnun á því hvað af hugmyndum hennar og fræðilegum niðurstöðum hafi leitað í rit annarra fræðimanna hefur þó ekki verið gerð og því er ekki hægt að fullyrða hvort áhrif hennar séu meiri en virðist" [*Based on prevailing ideas about mental illnesses, such conditions are often seen as diminishing a person's contributions in any field. Whether this was the reason or other factors caused Björg's work and legacy to fall into obscurity remains uncertain. In any case, Björg's scholarship was swept aside from the academic field and likely had limited influence, except for her writings on nutrition aimed at Icelandic housewives. A systematic examination of how her ideas and findings may have influenced the writings of other scholars has not been conducted, making it impossible to determine whether her impact is greater than it appears*] (online publication).

31 Based on the corpora of Modern Icelandic, the term "bætiefni" (lit. added material/element) is commonly not used to specifically replace "vitamins", but

to refer to the substances added to fortify food and medicines or contained in them. Ragnarsson (1979) provides one of the first translations of "bætiefni" as "essential nutrient".

32 As it appears in *Á víð og dreif. Ritgerðir* (1947) by Árni Pálsson.

33 Few examples show as well the impact of media on marketing successfully the substances imperceptible to the naked eye, as vitamins. Sounding scientific mattered for the producers of vitaminic supplements and nutrition was one of the first fronts in which commercial narratives were wrapped in material that could discursively sound scientific. In fact, even "before vitamins were isolated or could be synthesized, alert entrepreneurs built a vitamin industry on the promises of vitamin research" (Apple, 1996, p. 8).

34 Vitamin injection as an expression appears in books, newspapers and magazines around the 1940s and 1950s in the United States, "but we wonder if the last whistle acts as a vitamin injection to him at quitting time" (*Toledo Union Journal*, 1943), "something that promises to inject much-needed vitamins into the depressed textile industry of Lancashire" (*Evening Star*, 14 April 1952), "Mrs. Eisenhower's frank enthusiasm for average priced American ready-to-wear has worked like a vitamin injection on vast U.S. garment industry" (Roe, 1953, p. 12); in The Netherlands, "de Nationale Woningraad, door het samenzijn van vele leden een „vitamin-injectie" ontvangt die kracht geeft tot het verrichten van haar gewichtige taak" [the National Housing Council, through the gathering of many members, receives a "vitamin injection" that provides strength to carry out its important task] (*De woningbouwvereniging*, 1950, p. 40), "een muzikale vitamine-injectie" [a musical vitamin injection] (*Het Dagblad*, 18 September 1946), "een (kleine) verbetering aan de oude lijn (Amsterdam-Rozendaal), de ruggegraat van NS, die af en toe een vitamine-injectie nodig heeft" [a (small) improvement to the old line (Amsterdam-Rozendaal), the backbone of NS, which occasionally needs a vitamin injection] (*Het Parool*, 26 July 1955); in Sweden, "Den verkar som en vitamininjektion och lär ofta leda" [It acts like a vitamin injection and often tends to lead] (*Arbetet*, 15 September 1953); "som en vitamininjektion på min son sade fru Nilsson" [like a vitamin injection for my son, said Mrs. Nilsson] (*Aftonbladet*, 21 April 1954); and in the United Kingdom, "the impact of two pungent comments — one American and one German — on recognizable realities of our time is as stimulating as a vitamin injection" (Lockhart, 1950). The expression is found in several other languages, including Spanish.

35 Over the production of vitamins, Eggersdorfer (2012) distinguishes notes roughly four phases, "The pioneering days from 1930 to 1950; – the scaling-up and engineering phase from 1950 to 1970; – the period of worldwide production plants followed by consolidation from 1970 to 1990; and – the period of the rise of new technologies from the 1990s to the present day" (p. 12965).

36 To function properly, humans must obtain from their diet essential amino acids, certain type of fatty acids and minerals.

37 Identified in Icelandic Risamálheildin Corpus (2019).

38 There are only three examples documented in the corpora (1976, 1987, 1997).

39 Before being renamed Ciba in 1884, the chemical company was called Gesellschaft für Chemische Industrie Basel after it was bought by Bindschedler & Busch in 1874 (HLS, 2007).

40 See for instance *Periódico Oficial del Estado de Guerrero*. 16 April 1927.

41 *Gaceta Médica de México*. 1 August 1932.

42 It contained a barbiturate derivative in combination with amino phenazone, a discovery credited to Ernst Grether and Ernst Preiswerk.

43 Allobarbitals, such as Cibalgina (Cibalgine in English) were also used as antipyretic and anticonvulsant.

44 Gutiérrez Rodilla (2010) asserts that Alonso y de los Ruices de Fontecha (1606) based his midwifery treatise in great part on vocabularies related to materia media presumably derived from Arabic oral tradition. Etymologically, it is a Greek reintroduced loanword, ὀδονταλγία (odontalgía), derived from ὀδόν-τος (odōn) (odóntos/ tooth) and ἄλγος (álgos/ pain), traceable back to Dioscorides.

45 In the 18th century, it appears in translated treatises (e.g., *Sistemas de cirugía* by Benjamin Bell, translated by Santiago García, 1788) and general compendia (e.g., *Compendio de la medicina, ó Medicina practica* by Juan Manuel Venegas, 1788).

46 Following the merger of J.R. Geigy and Ciba AG in 1970, the pharmaceutical company received the name Ciba-Geigy until it merged with Sandoz in 1996 to form Novartis AG.

47 A manual written by Antonio Ruiz and published by Penthalon in Madrid in 1993.

48 Montero's novel was first published by Editorial Debate in Madrid in 1988 whereas Plaza y Janés released *El Alpe d'Huez* a year after García Sánchez completed the manuscript in 1995.

49 Rosa Montero Official Site. 2023.

50 The commercialization of the first approved oral contraceptive pill started in 1957. Enovid (USA)/Enavid (UK), as it was called by its developer Gregory Pincus from Searle & Company was a combined preparation of mestranol and norethynodrel, a synthetic progestogen first synthesized by Frank Colton in 1952 and preceded by the synthesis of norethindone/norethisterone by Luis Ernesto Miramontes Cárdenas, Carl Djerassi and George Rosenkratz in Mexico City in 1951. The medical trials for Enovid were set up in Latin America (Puerto Rico, Haiti and Mexico) in 1956 (Galán, 2010).

51 Similar, for instance, to Anovial 21, a 21-tablet regime, known in other countries as Anovlar and introduced in Spain in 1964 (Rodríguez-Ocaña *et al.*, 2012).

52 Officially in January 1974 by means of Ley General de Población (General Law of Population) (Secretaría de Salud, 2014).

53 Probably as it was the case in most of the world. White Junod and Marks (2002) assert that from "the time of the first clinical trials to the present, nearly 200 million women have swallowed various formulations of the contraceptive pill, making it one of the most widely consumed class of drugs in the world" (p. 117).

54 On rhetorical devices within the publicity of medicines, see Møllebæk (2020).

5 Linguistic By-Products of Pharmacy

This last chapter focuses on the adoption, appropriation, and use of pharmaceutical and medicine-related lexicon in spoken language. As I show, the commercialization of medicines has had a linguistic impact on Icelandic and Spanish speakers at various levels and in unsuspected ways. Surpassing the boundaries of prescriptiveness, oral and written references to brand names as well as pharmaceutical and scientific terminology, illustrate how the legal, economic, cultural, and biosocial aspects of medicines, as substances for the care and cure of the body, are adopted sociolinguistically. As the excerpts of the corpora and transcribed interviews show, the linguistic implications transcend scientific considerations and regulatory protocols. Even if, as I have shown, the language of pharmacy is logistically managed, how speakers—actually— speak escapes any predetermined schematization. The first section, "The Sounds of Sexual Pharmacy", continues the discussion about contraceptive pills started in the previous chapter, adding to the debate the biocultural and sociolinguistic traces of sildenafil (Viagra) in the pharmacological imaginary and linguistic output. This discussion is followed by an analysis of colloquial terms used to refer to mind-altering substances in Icelandic and briefly in Spanish. The section "Psychoactive Substances Slang" delves into an analysis of "slettur, bannorð og annað utangarðsmál" [*foreign loanwords, banned words and other peripheral use of language*)], (Árnason *et al.*, 1982) as they were first compiled and published in an argot dictionary in the 1980s. At last, the chapter ends with the section "Pharmaceutical Aural and Oral Side Effects", where I discuss the expressions, *Ic.* "ozempicandlit" and *Sp.* "cara ozempic" (Ozempic face), popularized by social media commentators in early 2022 to label the aftereffect of weight-management pharmacological preparations on the faces of patients/users. Through the revision of identified patterns in spoken language, the chapter conclusively reflects on how medicinal jargon and branding inform the adoption of pharmacy-related language and soundscapes in the 21st century Iceland and Spanish-speaking world.

DOI: 10.4324/9781003601654-6

5.1 The Sounds of Sexual Pharmacy

Pharmacy is central in facilitating access to reproductive and sexual health commodities, many of which users have relabeled colloquially, renamed euphemistically or to which they have reassigned new meanings or affixed new references. The language of sexual pharmacy evolves rapidly. Most probably because it alludes to areas of human health that are often subject to moral, political, and religious debates and controversies. Nevertheless, based on historic corpora and compared to previous decades, in the 21st century, it is possible to establish that there is less stigma, taboo and reservation about fertility control and sexual intercourse in the Spanish-speaking world. While the first part of this section revolves around the "morning-after pill" (*Sp.* pastilla del día después) (Figure 5.1), the second part reviews the popularity of Viagra as a linguistic and pragmatic-semantic reference in spoken Spanish.

As the interview on "la píldora" (the pill) in the previous chapter illustrates, contraceptives are embedded in processes with social ramifications at various levels. Many of which have to do with language. Firstly, as the informant accentuates, sexual education revolves

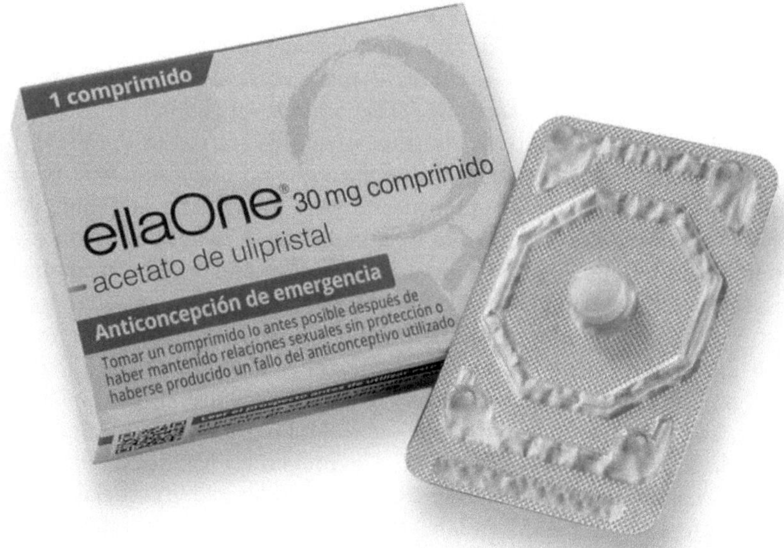

Figure 5.1 EllaOne brand morning-after pill blister pack. HRA Pharma.

Source: Photographer: Julio Monterrubio (2017)

around "hablarlo" [*speaking thing out*], according to her, preferably "con crudeza" [crudely] and "no adornando la píldora como digo yo" [*not embellishing the pill as I say*] (PRESEEA LHAB_M32_067). In her view, it is a matter of finding the right words. Secondly, from her interview and the next natural language excerpts, it matters that potential users/patients can access any means needed to safeguard their well-being and find venues to talk openly about their concerns. Thirdly, there is a complex interrelation between information, marketing and breaking down linguistic taboos, as the following excerpt illustrates (see Table 5.1).

Table 5.1 PRESEEA Oral Corpus excerpt on "the morning-after pill" in Spain

SEVI_H11_002.mp3 PRESEEA Corpus 2017 Seville, Spain Informant: Female, 31 years old	
I: muy delicado / pero que lo veo muy mal muy mal muy mal / el tema de los abortos / porque<alargamiento/> / hoy en día se queda embarazada el que quiere	I: very delicate / but I see it as very bad, very bad, very bad / the issue of abortions / because<lengthening/> / nowadays, whoever wants to get pregnant
E: uhum	E: um
I: porque hay muchísimos<alargamiento/> anticonceptivos / muchísimo información<alargamiento/> / muchísimas cosas que pueda evitarlo / hoy en día / casi todos los abortos que hay / son / sinceramente / chavalas jóvenes / que van a lo que van	I: because there are so many <lengthening/> contraceptives / lots of information <lengthening/> / so many things that can prevent it / today / almost all the abortions that exist / are / honestly / young girls / who go for what they go for
E: uhum	E: um
I: <tiempo = "34:37"/> el aquí te pillo aquí te mato <risas = "E"/> y<alargamiento/> <risas = "I"/> / a ver si me quedo / voy a tal sitio me lo quito / antes te daban la pasti<palabra_cortada/> yo me acuerdo que te daban la pastilla<alargamiento/> del día después	I: <time = "34:37"/> the "I'll catch you here, I'll kill you here" <laughs = "E"/> and<lengthening/> <laughs = "I"/> / let's see if I get pregnant / I'll go to a certain place and get rid of it / before they used to give you the morn<word_cut_off/> I remember they gave you the morning<lengthening/> after pill
E: uhum	E: um

(*Continued*)

Table 5.1 (Continued)

I: ya no te la dan	I: they don't give it to you anymore
E: ¿ah no? / ¿ya no te la dan?	I: oh no? / they don't give it to you anymore?
I: ya no te la dan / por culpa de ese tema / porque las niñas se creían que las pastillas era como las pastillas juanolas	I: they don't give it to you anymore / because of that issue / because girls thought the pills were like Juanolas lozenges
E: ¡madre mía! <risas = "E"/>	I: oh my gosh! <chuckles = "E"/>
I: yo me harto / me la tomo / yo me harto y yo me la tomo / y los preservativos para hacer el globito nada más / entonces / y la <vacilación/> la<alargamiento/> Seguridad Social habrá pensado / esto es un dineral que nos estamos dejando aquí / los niños no ponen remedio / porque existe los preservativos / que te puede pasar que se te puede romper / cualquier cosa / te la tomas / pero que esa gente lo tenía como si fuesen gusanitos / pim pim para adentro / pim pim para adentro / y así pues tuvieron que quitar y cortarle el rollo	I: I'm fed up / I take it / I'm fed up, and I take it / and the condoms, just to make little balloons, nothing else / so / and the <hesitation/> the<lengthening/> Social Security must have thought / this is a lot of money we're spending here / kids don't take precautions / because condoms exist / and sure, they can break / anything can happen / you take the pill / but these people treated it like it was candy worms / pim pim inside / pim pim inside / and so they had to stop and cut the fun
E: claro	E: of course
I: a ver si así se<alargamiento/> centran un poco más en esas cosas	: let's see if that way they<lengthening/> focus a little more on those things
E: ¿pero tú crees que todavía hace falta más educación en un <vacilación/> / <simultáneo> para los jóvenes? no ¿no? </simultáneo>	E: but do you think more education is still needed in a <hesitation/> / <simultaneous> for young people? No, right? </simultaneous>
I: <simultáneo> los jóvenes<alargamiento/> ya <vacilación/> </simultáneo> creo que ya está <vacilación/> / más información no pueden darle ya / si es que ya veo <vacilación/> creo que lo saben todo	I: <simultaneous> young people<lengthening/> already <hesitation/> </simultaneous> I think it's enough <hesitation/> / they can't give them more information at this point / the way I see it <hesitation/> I think they know everything

(*Continued*)

Table 5.1 (Continued)

E: uhum	E: um
I: ya lo hacen porque quieren hacerlo	I: they already do it because they want to do it
E: ya	E: right
I: se quedan porque quieren / porque ya<alargamiento/> / ¿qué más información? / ¿qué<alargamiento/> publicidad? / ¿qué<alargamiento/> cosas se dan en el colegio? / tus padres <vacilación> tu familia te informa / todo / es que ya / no sé qué más pueden sacar / ya se quedan porque quieren	I: they get pregnant because they want to / because by now<lengthening/> / what more information? / what<lengthening/> advertising? / what<lengthening/> things do they teach you at school? / your parents <hesitation> your family informs you / everything / it's already enough / I don't know what else they can come up with / they get pregnant because they want to
E: ya	E: right
I: porque pueden poner muchísimos remedios	I: because they could use many ways to prevent it
E: <simultáneo> uhum </simultáneo>	E: <simultaneous> uhum </simultaneous>
I: <simultáneo> están </simultáneo> las pastillas anticonceptivas / que te las estás tomando y no te quedas	I: <simultaneous> there are </simultaneous> contraceptive pills / you take them and you don't get pregnant
E: uhum	E: um
I: pero es más fácil<alargamiento/> no tomarte nada / y ahí al grano	I: but it's easier <lengthening/> not to take anything / and get straight to it
E: claro / uhum / muy bien / ¿y de la anorexia qué piensas?	E: sure / uhum / very good / and what do you think about anorexia?

The informant does not approve of abortion as she considers that there are enough birth control methods. There are many "anticonceptivos" (contraceptives) and "muchísimas cosas" (many many things) to avoid being pregnant, besides a lot of information. The "pastilla del día despúes" (morning-after pill), according to her, is not an option anymore due to misuse and abuse of the health system. The comparison between emergency contraceptive pills and Juanola liquorice pills is somehow sarcastic, slightly humoristic, and yet somehow paternalistic. These small lozenges have enjoyed renown popularity throughout Spain since the early 20th century and are presumed to contain, in a small quantity, a secret medicinal formula, not publicly disclosed, but celebrated for its artisanal origin. The link between candy and medicines does suggest the perceived notion that pharmaceuticals are (were) widely available. The remark on "preservativos" (condoms) as raw material for toy balloons confirms the ludic sense in which sexual health commodities are discussed. The action of taking medicines is described through the standard reflexive form of "tomar"> "tomarse" and the onomatopoeic "pim pim", denoting the noise when a task is completed, a bell rung or a sip swallowed. In terms of information, the interviewee thinks young people receive enough sexual education. In fact, her view is that there is a bit of saturation on the topic. There are plenty of "remedios" (remedies), but the easiest thing, said ironically, is "no tomarte nada" (not take anything), which possibly confirms, she refers to taking medicines orally. The use of "remedio" implies a method to mitigate a condition, but as the informant notes, birth control revolves around prevention. From this angle, labeling contraceptives "remedios" showcases the broad spectrum that pharmacy is assumed to encompass in terms of efficacy. Similarly, the remark underlines the casual speech about conception control methods, compared to previous decades. From the linguistic taboo perspective in regard to emergency contraceptive pills, as the next excerpt illustrates, media outlets are again mentioned as key players in changes in public debates on sexual pharmacy practices (see Table 5.2).

The informant reflects retrospectively on the unusual attention that the emergency contraceptive pharmaceuticals received in the 1980s, particularly in mass media outlets: "hasta en las noticias la sacaban" [*it was even on the news*]. She goes on to mention four names by which the pills have been known, "pastilla del día después" (day-after pill), "pastilla de emergencia" (emergency pill), "anticonceptivo de emergencia" (emergency contraceptive) and "pastilla de anticoncepción de emergencia" (emergency contraceptive pill), while commenting extensively on the users. Unexpectedly, for the informant, the morning- after pill was common among educated women who did not have any other birth control contingent plan, at odds with a preconceived link between education and preventive health measurements. From a linguistic perspective, two aspects stand out. On one hand, no dominant brand name became the main referent for the emergency contraceptive pills. The commercial names most found in the NOW Corpus include ellaOne (23 entries) and NorLevo (20 entries). On the other hand, when mentioned, the names commonly found

Table 5.2 PRESEEA Oral Corpus excerpt on “the morning-after pill” in Mexico

MEXI_M13_012.wav PRESEEA Corpus 2005 Mexico City, Mexico Informant: Woman, 21 years old	
I: <tiempo = “35:28”/> pero ¿sabes qué? // la otra vez estábamos haciendo // un<alargamiento/> // un estudio // no tenía nada que ver con el virus del papiloma pero era más o menos algo relacionado / era de cuando salió todo el bum de / la pastilla del día después / que hasta en las noticias la sacaban / y <cita> ay / la pastilla del día después / o la pastilla de emergencia // el anticonceptivo de emergencia </cita> // entonces / empezamos a revisar artículos desde el<alargamiento/> / desde que se inventó casi casi la pastilla de emergencia / que fue como en lo <palabra_cortada/> lo <palabra_cortada/> los ochenta / por ahí así // y y es impresionante porque ves los artículos // y la mayoría de los testimonios que hay / son personas que / que están educadas / y esas personas son las que consumen más la / la pastilla de anticoncepción de emergencia / y tú dices / <cita> ¿cómo puede ser? </cita> o sea // es una persona educada / que tiene un nivel // cultural / adecuado / normal en donde pues se puede meter a Internet / o puede leer un libro / y sí tiene / y y lo lee / o por lo menos sabe / y ellas son las que más lo utilizan / a las personas que no / las personas que no saben como que se cuidan más // entonces tú dices / <cita> ¿pues qué está pasando? </cita> / entonces ¿por qué está todo al revés? //	I: <time = "35:28"/> but you know what? // the other day we were doing // a<lengthening/> // a study // it had nothing to do with the HPV virus, but it was somewhat related / it was from when the whole boom of / the morning-after pill came out / and it was even on the news / and <quote> oh / the morning-after pill / or the emergency pill // the emergency contraceptive </quote> // so / we started reviewing articles from<lengthening/> / from when the emergency contraceptive pill was almost invented / which was like in the<word_cut_off/> the<word_cut_off/> the eighties / something like that // and it's impressive because you see the articles // and most of the testimonies there are / are from people who / who are educated / and those people are the ones who use the / the emergency contraceptive pill the most / and you say / <quote> how can that be? </quote> I mean // it's an educated person / who has a level // of culture / adequate / normal where they can go online / or read a book / and they do / and they read it / or at least they know / and they're the ones who use it the most / while the people who don't / the people who don't know seem to take better care of themselves // so you say / <quote> what's going on? </quote> / so why is everything upside down? //

in spoken language are descriptive and devoid of the linguistic opacity that characterized sexual pharmacy medicines used in the past. As Gonsalves (2017) asserts, especially in the case of young people, "pharmacists and pharmacy personnel should be recognized as important complements to the role that physicians and other medical practitioners play in the delivery of SRH [sexual and reproductive health] services" (360). The case of the emergency contraceptive pill, as experienced by the Spanish speakers in Seville and Mexico City indicates the users were not, even if momentarily, stigmatized. Language sits at the core of sexual pharmacy, nonetheless, because users/patients need to overcome any communicative hurdles to consult a physician, discuss the topic with their partners and collect the medicine at the pharmacist. For some, even searching information online can be an uncomfortable experience, for instance, in case of erectile dysfunction[1].

Few medicines are more evidently and paradoxically defined by the intersectional axes of gender, geopolitics, patriarchy, pharmaceuticalization, illegality, and sociolinguistics as Viagra. Few medicines exist in as many regulated and unregulated brands, causing in the process the deployment of a language at a granular level not commonly needed for other types of pharmaceutical. Viagra makes people talk, people talk about Viagra and as the entries from the corpora show, in the process, the "blue pill" prompts one of the best study cases for the sociolinguistic impact of pharmacy on natural languages. Nonetheless in the case of a medicine that is used at the core of sexual pharmacy in a context heavily informed by imaginaries about masculinity, gender roles, taboo, and performance. The language about Viagra bears the earmarks of euphemisms and secrecy, but also the pharmaceuticalization of culture at a sociolinguistic level. Not all actors involved in the fabrication, circulation, consumption, and illegal trafficking of Viagra speak the same language, both literally and figuratively. Erectile dysfunction, colloquially known as impotency, is not an easy topic. It requires words that moderate and reflect the nature of this medicine at its core. Transculturally, as much as trans-linguistically, some clear-cut differences have been identified.[2] The language of sex is not the language of love. "In the process, there is an acute danger that anything less than a hyperperformance will be pathologized" (Weeks, 2011, p. 220). The use of Viagra as a synonym for vitality, potency, and thrust are not all related to romance, intimacy, or care, as the examples from NOW Corpus illustrate:

1 Viagra? –Por Dios que no, pregúntale a mi mujer. Mi viagra es mi pescado... todo es concentración (Peru 2012).
[*Viagra? –For God's sake, no, ask my wife. My Viagra is my fish... it's all about focus*]

2 La sandía se ha convertido en el viagra de los pobres, pues aumenta el deseo sexual (Ecuador 2011).
[*Watermelon has become the viagra of the poor, as it increases sexual desire*]

3 En mi tierra se dice que el café es el viagra de los pobres (Peru 2018).
[*In my hometown, they say coffee is the poor man's Viagra*]

What Viagra came to mean and its associations in the therapeutic imaginary, at a global scale, were by-products of the mobilization of this medicine throughout the world. Across Latin America, the visibility of Viagra across the sexual pharmacy landscape during the first and second decades of the 21st century resulted in significant profits for Pfizer. Once allowed, the interest in sildenafil also materialized in the involvement of local pharmaceutical companies in the production of generic versions under various brand names.[3] As doctors in Ecuador noticed between 2013 and 2015, they were reluctant to prescribe Viagra to their patients to avoid "recomendar algo de compra popular" [*recommending something widely accessible*] (Vanoni and Burgos, 2018, p. 32). Many of them "ven a nuestro producto como OTC (Over the counter)" [*perceive our product as an over-the-counter medicine*] (p. 32). Johnson (2016) rightly notes that "(D)rugs can both create new and refract with existing stereotypical images of a patient, images which carry markers of class, race and sexuality" and which can "lead to new laws and policies to regulate the practices of both doctors and patients" (p. 106). Images, markers, sociocultural and political implications that emphatically redefine the meaning of a name.

There is nothing neutral about any discussion in regard to Viagra or its use as a reference. Language is never dispassionate, but certainly not when one mentions a medicine used for the treatment of erectile dysfunction because of its taboo-defined connotation. The linguistic identity of this medicine has been changed by its place within the clinical, political, social, and more importantly, personal context in which it exists. Viagra can be regarded as a cultural revolution, one inscribed in shelves of pharmacies, literary metaphors in newspapers and on the walls of cities across the world (Figure 5.2). Chances are, most readers are more familiar with the brand name Viagra than with some of the most widely sold medicines on the planet, such as Humira, Keytruda, Eliquis, Revlimid, Imbruvica, Stelara, Eylea, Biktarvy, or Opdivo, to mention a few. All of which are names also coined to denote formulas logistically managed at a scientific, linguistic, and commercial level. Viagra is an interesting case in terms of language for various reasons. One of them is that it has mutated and entered other dimensions, showing the fertile capacity of language to register changes

Figure 5.2 Signage with Viagra Ads, Ensenada, Baja California Norte (Mexico) (2012).

Source: Photographer: Adam Jones: https://commons.wikimedia.org/wiki/File:Signage_with_Viagra_Ads_-_Ensenada_-_Baja_California_Norte_-_Mexico_(6776103046).jpg

and additions in epistemic and social systems. For instance, the plethora of falsified medicines with names inspired in Viagra sold in the black market, such as Furunbao, Kasutra and VisexPro (AEMPS, 2015). Street names of tadalafil, presumably in a substandard formula or combined with substandard components that have prompted the authorities to issue warnings. From a sociolinguistic perspective, it matters to reflect on the appeal of Furunbao, for instance, a product dubbed as "La viagra para jóvenes de veinte a treinta" [*Viagra for young people aged twenty to thirty*] (De la Cal, 2015).[4] A formula authorities doubt is truly made based on traditional Chinese medicinal knowledge,[5] which is extensive and multifaceted.[6]

The language of pharmacy is logistically planned. Whereas the discovery of Viagra stemmed from an episode of serendipity, linguistically it is a toponym derived from "vigor" and "Niagara" (i.e., Niagara Falls).[7] There are no apparent markers or references in the brand name itself, as opposed to its chemical formula (sildenafil). Yet, it has become a word found hundreds of times in the different Spanish and Icelandic corpora and print media consulted for this book. It was after all managed and coined to reflect the popularity expected by its creators, Pfizer, anticipated it would tap into a specific *market*. The commercial aspect of medicines is indeed an important factor in the linguistic production surrounding pharmaceuticals. Viagra might fall into the category of brand names that allude to abstract concepts (i.e., vigor) and topographical references (i.e., Niagara). But it was not created to pay homage to a large freshwater ecosystem. It emerged as a proper noun expected to encapsulate an expected degree of memorability, musicality, allure, and distinctness. Viagra is also a portmanteau designed to create two linguistic identities for sildenafil used for two different conditions. After all, "(O)n peut effectivement facilement imaginer la gêne d'un patient souffrant d'hypertension pulmonaire obligé de demander une boîte de Viagra® à son pharmacien" [*one can indeed easily imagine the embarrassment of a patient suffering from pulmonary hypertension forced to ask his pharmacist for a box of Viagra*®] (Faure, 2014, p. 85). There is a reason why sildenafil is marketed under two different names (Revatio and Viagra). Drug names are regulated, requiring approval for new drugs to prevent confusion with existing medications. Part of the process includes enforcing packaging regulations that ensure clear rules regarding adulterants and prohibit fraudulent or misleading labeling. Beyond the legal and technical aspects of a registered name, the discussion surrounding sildenafil and its efficacy among (potential) consumers, particularly on social media platforms, carries a sociolinguistic dimension. One that illustrates how the language of pharmacy is adapted by speakers and operates as an interface between the materialization of science and society.

Medicalization of sexual intercourse in one way or another has gone through various stages across time and space. Beyond the obvious cultural and biosocial implications of human procreation, most cultures have a repertoire of practices, norms, and ideas about copulation and intimacy. Attitudes about imagined and predetermined gender roles and behaviors explain much of the language deployed about Viagra as a remedy, cure, intervention, prosthetic aid, confidence booster, and a type of aphrodisiac. "According to this discourse, a person (or couple) can maintain a successful sex life, as long as penetration is possible, that is with the help of Viagra" (Johnson and Åsberg, 2016, p. 78).[8] In this sense, erasing any erectile disturbance is assumed to secure pleasant, fulfilling moments. Thus, talking about sildenafil and analog drugs is in fact discussing the perceived role and importance of biological males in sexual intercourse. A medicine with similar effect on female patients, even before being available, was widely announced as "feminine viagra":

4 Se de un tratamiento lógico seguro y eficaz "La aprobación de la viagra femenina, el pasado 21 de junio, brinda a las mujeres otra opción (Nicaragua 2019).
[*Safe and effective logical treatment. The approval of female viagra, on June 21, gives women another option*]

5 Que en 2015 se permitiera la comercialización de una píldora también conocida como la "Viagra femenina" (Venezuela 2019).
[*That in 2015 the commercialization of a pill also known as the "female Viagra" was allowed*]

Names of brands for erectile dysfunction medications respond to the set of norms, values, and assumptions that inform gender roles and sexual practices.[9] The linguistic implications of this set of beliefs are easily identified in the above-mentioned examples and many other entries in the corpora.

In terms of morphology, the grammatical gender of "viagra" (when applicable) seems to vary between countries, geographical regions and between speakers,[10] as the examples from the transcribed examples from the same country and city illustrate:

6 Pero conseguir lo es toda una odisea # ¡TOME NOTA! # El "viagra" de el Himalaya se convirtió en uno de los alimentos más caros (Panama 2019).
[*But getting it is quite an odyssey # TAKE NOTE! # The "viagra" of the Himalayas became one of the most expensive foods*]

7 A el día de hoy, casi 18 años después de el lanzamiento de la viagra, sigue siendo un tema tabú, que no se maneja (Panama 2019).
[*To this day, almost 18 years after the launch of viagra, it remains a taboo subject, which is not dealt with*]

For comparison, the same discrepancy in grammatical gender is found in two entries from two Colombian sources with same geographical origin and from the same year and month with two weeks in between their day of transcription:

8 Esta planta milenaria incluso aumentan la libido, según sus defensores: "Es la Viagra de las mujeres ", dicen estas jóvenes a pura sonrisa (Colombia 2019).
[*This ancient plant even boosts libido, according to its advocates: "It's the Viagra for women," say these young women with broad smiles*]

9 entre 2013 y 2015), lograron que sus pepas se vendieran más que el viagra y los posicionaran como la tercera familia más rica de Estados Unidos (Colombia 2019).
[*between 2013 and 2015), they managed to sell their seeds more than Viagra, positioning themselves as the third richest family in the United States*]

As for the plural form, in spoken language, Viagra is used in plural, as the following examples attest. While in the second case, "viagra" is a feminine noun, in the first one is a masculine noun that agrees in number and gender with the verb and past participle.

10 Para formar parte de la información Hay viagras alternativos que están hechos de hierbas y de vitaminas (Venezuela 2012).
[*To be part of the information There are alternative viagras that are made from herbs and vitamins*]

11 Registros domiciliarios en Barcelona los agentes intervinieron ampollas de popper, pastillas de éxtasis, viagras falsificadas, resguardos de envíos a través de empresas de paquetería y ordenadores portátiles (Spain 2012).
[*Home searches in Barcelona led agents to seize ampoules of poppers, ecstasy pills, counterfeit Viagra, shipping receipts from courier companies, and laptops*]

The coinage of various generic medicines, colloquialisms, euphemistic references, and comparisons in the aftermath of the development of Viagra, showcases the logistic management of pharmaceutical language. The branding strategies behind the erectile dysfunction medicine have "succeeded so thoroughly in rationalizing the idea of sexual correction and enhancement through pills" (Tiefer, 2006, p. 287). Narrating the benefits of Viagra involves the deployment of language that appeals to the public and personal understanding of how a pharmacological preparation embodies health management, commerce, and even cultural legacy, as the entries on *palwen* indicate:

12 Está el viagra marino, el viagra mapuche y el viagra farmacéutico, los cuales se venden libremente, generando actualmente (Chile 2013).
[*There is the marine viagra, the mapuche viagra, and the pharmaceutical viagra, which are sold freely, currently generating*]

13 Los magallánicos de todas las edades han solicitado harto el viagra mapuche, desde jóvenes hasta caballeros de edad (Chile 2013).

[*Magellan residents of all ages have eagerly requested Mapuche Viagra, from young people to elderly gentlemen*]

14 Producto que más se vende, afirma que es el viagra mapuche, ya que a su tienda llegan muchos señores y jóvenes con ese (Chile 2013).
[*The product that sells the most, he claims, is the Mapuche viagra, as many gentlemen and young people come to his store for it*]

Palwen is a Mapuche remedy sold as an aphrodisiac, colloquially called "Mapuche Viagra" and which enjoys great popularity in traditional medicine pharmacies in Chile (Cancino, 2017).[11]

15 El efecto dura 72 horas. El caso más curioso quizá sea el de el Viagra mapuche, denominado Palwén, que aprovecha la sabiduría de los recios indígenas que vivieron (Venezuela 2012).
[*The effect lasts 72 hours. Perhaps the most curious case is that of the Mapuche Viagra, called Palwén, which draws upon the wisdom of the resilient Indigenous people who lived...*]

In this context, the proper noun "viagra" acquires the pragmatic-semantical function of "erectile dysfunction medication", aphrodisiac and (sexual) stimulant. From this perspective, *palwen* is described through a categorization of a type of substance that can be used to enhance and reinvigorate libido and energy levels, regardless of the biological gender of the user.

In Foucauldian terms, the history of sexuality is a history of the dominant discourses that reflect "the necessities of their operation, the tactics they employ, the effects of power which underlie them and which they transmit" (Foucault, 1987 [1976], p. 68f). Nonetheless, in the case of Viagra, because of the type of "hyperbolic language" that is deployed every time there is a medical novelty "about erections" (Tiefer, 2006, p. 281). Pharmacy has created a link between sex and performance by technologizing ways to avoid perceived failure (Weeks, 2011) and hence, desired expectations. Deconstructing elements identified in the promotional material for sildenafil "reveals the interconnectedness of language and ideological presuppositions on the part of the pharmaceutical industries that reflect and reproduce heteronormative ideas and practices concerning masculinity" (Simon-Maeda 2008, p. 27). The conversation around Viagra elicits the identarian character of a pill developed to alleviate dysfunctional erection and the dominant discourse on the topic, mainly from the perspective of one biological gender. The vocabularies used to refer to "impotency" glean vivid images of shame, deviant performance, aging, and diminished virility. Manliness is questioned by the need of a pharmacological artifact in the context of sexual intercourse.

But at the same time, rhetorically, Viagra confirms the power of potential users, precisely by reminding them how much their identity aligns with the results they could imaginably obtain from taking the medication. As examples from the written and oral natural language corpora show, as a term, "viagra" is used as a synonym of energy, intensity, and force, but also as a category of medication that serves as benchmark for potency-enhancers and stimulants, even in unrelated contexts. As opposed to the language used in regard to the emergency contraceptive pill, the various terms applied to refer to the "blue pill" seem to conceal a self-righteous type of dysphemism. Semantically and pragmatically, time is left out of the lexicon used in the context of erectile dysfunction, as opposed to the morning-after pill. Even if, chronological considerations do apply to both. "La pastilla de emergencia" (emergency pill) could be easily used in spoken Spanish to refer to the tempo-pharmacological dimension of both medicines, but it is not the case. Contraception, unquestionably, demands immediate attention. However, through language, the sequential and consequential applicability of a medicine, can be emphasized or be tactically understated. In either case, what does apply to both pharmaceutical breakthroughs are the role of pharmacy in upholding a commercial-medical connection that legitimizes the commodification of medicines. One that, in the context of controversial practices, prompts the emergence of new lexicons and registers, as is the case of slang surrounding psychoactive substances.

5.2 Psychoactive Substances Slang

This section reviews the language surrounding "pleasurable (hedonic)" drugs, considered so because they "are consumed because people choose to, and not because they are advised to by their doctor" (Rang *et al.*, 2012, p. 592). Although historically, many of what we call today "drugs" with the connotation of illegal psychoactive substances, were once part of the repertoire of substances administered to patients by their physician or local chemist. From the 1960s onwards, a classification with legal implications redrew nevertheless a special category designated to mind-altering, addictive drugs.[12] The beginning of a war on illegal narcotics that has caused millions of lives without effectively eradicating the major social implications that drug addiction has on the patients, their families, and communities. In the 21st century, a different approach has been to decriminalize or have the state taking control, at least partially, of the "drug industry". Meanwhile, scientific research on the potential applications of mind-altering substances medicinally is reshaping the perception about forbidden plants, mushrooms, and lianas.[13] Since the legalization or decriminalization of medical cannabis in over 40 countries, for instance, pharmacies have become points of circulation and administration of a substance normally purchased in illegal circuits. Even more revolutionizing, since 2017, pharmacies in Uruguay are also the authorized sale points of cannabis for recreational purposes. For a long time and still today, prohibition, stigma, and taboo surrounding recreational drugs, among other reasons, have resulted in linguistic creative ways of speaking about and referring to these substances. Language has registered the sociocultural traces of drugs in Icelandic and Spanish through the creation of whole vocabularies to designate objects and practices related to psychotropic agents. By revising entries from an Icelandic slang dictionary and excerpts from Spanish oral corpora, this section focuses on drugs as linguistic byproducts of pharmacy.

In the preface of *Slangur: Orðabók um slangur, slettur, bannorð og annað* (1982), editors Mörður Árnason, Svavar Sigmundsson and Örnólfur Thorsson note that four areas are marked specially. As the title indicates, the dictionary includes words commonly classified as slang, cant, jargon, calques, loanwords, and urban slang. While compiling the dictionary, the authors observed the large amount of Icelandic slang (*Ic.* slangur)[14] words related to "málfari sjómanna (S), málfari tengdu neyslu ólöglegra fíkniefna (F), úr íþróttamáli (Í) og máli um tónlist, fyrst og fremst rokktónlist (T)" [*language of sailors (S), language associated with the use of illegal drugs (F), language from sports terminology (Í), and language about music, primarily rock music (T)*] (p. vi). As well as a large lexicon related to "áfengisneyslu og ýmis stig ölvunar" [*alcohol consumption and various levels of intoxication*] (p. vi). An additional remark is added about the possibility of having included slang dating from 1960, in combination with newer lexicon, up to 1982. From the paratextual details, one can infer that many were involved in the compilation of the entries, examples about their meaning and in some cases, the history of the objects, practices, qualities, and activities for which each slang word stands.[15]

One of the drug-related verbs is "éta", which means "to eat" and is normally used for animals or colloquially for eating hastily, without manners or uncoordinatedly. As the entry reads: "**éta.** so. 1 (F) neyta fíkniefna" [*to eat. verb. 1 (F) to consume drugs*] (Árnason et 1982, p. 29), for instance "éta sýru" [*to consume acid*], referring to LSD. Once acid is consumed, users "fara upp" (go up) or "lyftast" (get lifted, get high). After which they "koma niður" (come down) or "lenda" (land), during a "niðurtúr" (down trip) that typically occurs between 5 and 9 hours after the substance has been consumed. Acid users ("sýrufrík", "sýruhaus", "sýrukóngur") might experience a series of undesired effects that include paranoia, anxiety, and fear, jointly labeled "baddtripp" or "bömmer"; an example that resonated with the note in the preface about the many loanwords from English in Icelandic slang. "Djönka sig" (to drug oneself), derived from "junkie", "dóp" (drugs) derived from "dope" and "pöff" (puff) are other examples thereof. Almost two decades before the publication of the dictionary, Groenke (1966) advocated for the importance of distinguishing between "the categories of 'genuine substandard' and 'artificial substandard' in Icelandic slang" (p. 225).[16] Foreign words, particularly anglicisms (Kvaran, 2004), can commonly be heard as slang in informal daily speech, but are avoided in formal points of reference for language, such as mass media and translations (Ragnarsdóttir, 1996). And are often underrepresented in material used for linguistic analysis, such as corpora and textbooks. Yet, slang, along with loanwords and informal language are among the first sources "where changes to the system will first appear" (Bjarnadóttir, 2012, p. 15).[17] Paradoxically, more knowledge about the informal lexicon of drugs might be the key to gaining insights about the consumption of illegal substances (Simpson *et al.*, 2018).

The appetite for pharmaceuticals used as recreational drugs explains the broad lexicon of terms to refer to medicines normally administered by a physician. "Læknadóp"(lit. medical drugs) is described as an amalgam of pharmaceuticals that can be used recreationally because of their effect on the brain and nervous system. On this term, Jóhannsson (2011) notes that, although "Sumum er illa við orðið læknadóp en öðrum finnst best að tala tæpitungulaust um hlutina" [*Some dislike the term medical dope, while others prefer to speak plainly about things*], the term simply refers to "vanabindandi lyf sem læknar hafa ávísað en er notað í öðrum tilgangi en til var ætlast, sem vímuefni eða dóp" [*addictive drugs prescribed by doctors but used for purposes other than intended, as intoxicants or recreational drugs*] (p. 519). As Table 5.3 shows, the Icelandic street terminology for pharmaceuticals (e.g., "barbar" (barbiturates)) in the 1980s was quite rich, bearing in mind that in Icelandic, as in any other language, "Eitt einkenni slanguryrða er einmitt að þau úreldast fljót" [*One of the characteristics of slang words is precisely that they become obsolete quickly*] (Kvaran, 2004b, n.p.).

On the words listed in *Slangur: Orðabók um slangur*, Kvaran (2004b) also notes that much "í þeirri bók er nú, rúmum tuttugu árum síðar, horfið í gleymsku og annað slangur er notað í dag" [*that book is now, over twenty years later, forgotten and different slang is used today*] (n.p.). An aspect to consider in the context of the substances listed in Table 5.3, due to the constant changes in the array of substances mishandled for their pharmacological effects. Based on Ármason *et al.* (1982), the table includes slang words for weight-management medicines that were also part of the repertoire of ab- and misused pharmaceuticals:

As for drugs derived from plants, some of the slang words include references to their places of origin or cultivation, such as the case of cannabis (e.g., "afgan", "indverji, "líbani/líbanon", "nepall", "marokkó",

Table 5.3 Slang terms for pharmaceuticals in Icelandic, based on Árnason et al. (1982)

Scientific name/pharmaceutical name	Icelandic slang word
Amyl nitrate	popp, poppís, poppers (*En.*)
Barbiturates	barbar
Diazepam	stessa, valla, Valla vin/viking
Ephedrine	effi
Hydrate of chloral	klór, klóral
Nitrazepam	moggadon, mókadon
Nitroglycerin	glussi, sprengipilla
Pentobarbital	mebbi, natríum, natan, kría
Phenmetrazine	prelli
Phentermine	míri, mírapoint
Valium	valla, Valla vin/viking

"pakistan" and "tyrki", referring respectively to Afghanistan, India, Lebanon, Nepal, Morocco, Pakistan, and Türkiye). The plethora of terms for "marihuana" (herbal cannabis) include: "grænmeti", "hey", "Jóna", "Maríanna", "María Anna", "María Jóna", "mæja", "pott" and "pottur", whereas the slang words for "hashish", the herb made by processing cannabis, include: "dúndur", "efni", "eitur", "gúmmí", "kúkur", "kvef", "lakkrís", "lyfjagras", "lyftiduft", "mold", "piff", "pöff", "sass", "sissí", "sjitt", "skítur", "smókur", "stoff", "stuð" and "stóff". Many of which are standard words for objects, such as, "grænmeti" (vegetable), "lakkrís" (liquorice) and "kúkur" (feces). There are also onomatopoeias, proper nouns, and calques from English.[18] There seems to be a difference in the slang used for substances that are perceived to be manufactured chemically and those that are assumed to be cultivated. Probably due to its relative popularity, considering it was and is illegal in Iceland to consume cannabis, according to the entries in *Orðabók um slangur*, the lexical creativity and specificity are quite impressive. To make this point, the next overview displays a list of adjectives, verbs, and nouns that were used in the context of cannabis in Icelandic slang in the 1980s (see Table 5.4).

Although some words have a different meaning in standard Icelandic, such as "lesa" (to read), "reyktur" (smoked) and "lyfjafræðingur" (pharmacist), many of the above-listed terms are slang words not found in dictionaries. In the case of verbs, most of them are reflexive (sig > oneself) and one appears in middle voice (miðmynd), which also alludes to an action affecting or involving oneself, in certain ways, comparable to passive voice in terms of meaning. Many of the adjectives build on the lexeme "steinn" (stone),[19] whereas four nouns are derived from "hass" (hashish) and "dóp" (dope).[20] As a substance, cannabis is articulated

Table 5.4 Cannabis-related slang based on Árnason et al. (1982, p. 47)

Verbs used to refer to the action of consuming cannabis (and possibly other drugs)	Verbs used to describe the activity of smuggling or selling cannabis (and possibly other drugs)	Adjectives to describe a consumer under the effect of cannabis (i.e., drugged, high, stoned)	Nouns to describe a consumer of cannabis (and possibly of other drugs)
dópa sig	díla	bergnuminn	dópari
eitra sig	pússa	dúndraður	dópisti
fá sér eina feita/grana	þjappa	grýttur	eiturhundur
		hassaður	hassari
fýra í (pípu)		hátt uppi	hassasti
fæfa sig		hæ	lyfjafræðingur
lesa		lyfjaður	svælari
lyfja sig		reyktur	
pöffa		upp á rönd	
reykja		sissí	
smika sig		skakkur	
smóka sig		smikaður	
stóna sig		smókaður	
stönka sig		steinaður	
stónkast		steindur	
stuðla sig		steingerður	
stöffa sig		steinrunninn	
		stónaður	
		stónd	
		stónkaður	

linguistically at various levels due to the gray areas in which it circulates in countries where it is prescribed as a medicine. That list does not include Iceland and this will probably keep reflecting in Icelandic slang.

In line with the distinction between pharmacy and pharmaceuticals, the language of pharmacy also includes the lexicon related to substances with pharmacological properties, regardless of their legal status, socio-cultural implications, and history. Starting in the 1600s, Duffull (2018) suggests that the role of the pharmacist had undergone four transition states: "(1) manufacturer and purveyor of medicinal products, (2) dispenser and manufacturer, (3) dispenser and clinical checker and (4) clinical service provider" (310). Turning to the Spanish-speaking world, the last phase is probably the state in which pharmacists in Uruguay are, based on the above-mentioned legalization of recreational cannabis and which stipulates that pharmacists administer them. Seven years before the legislation was approved, a Montevideo-based informant recalled a passage of his childhood in these terms (see Table 5.5).

Table 5.5 PRESEEA Oral Corpus excerpt on drugs in Pocitos, Uruguay

MONV_H33_033.mp3 PRESEEA Corpus 2010 Montevideo, Uruguay Informant: Male, 77 years old	
I.: dicen que esa zona es de porro / de droga y de delincuencia / pero no te puedo / afirmar / es un	I.: They say that this area is for joints / drugs and crime / but I can't / tell you / it's a
E.: sí	E: yes
I.: sólo un / un boca a boca	I.: just a / a word of mouth
E.: bueno yo cuando / cuando era chica veníamos mucho a la playa de Pocitos y yo me acuerdo que mi madre siempre decía que era / la más peligrosa de venir / porque era la que más gente robaba / en Pocitos / pero en la playa que no quiere decir que sea la gente del barrio en realidad	E.: well when / when I was a girl we used to come to the Pocitos beach a lot and I remember that my mother always said that it was / the most dangerous area to come / because it was the one where most people got mugged / in Pocitos / but on the beach that does not mean that it is actually the people of the neighborhood

The informant uses the term "porro" (joint) while describing the character of an area based on a substance being consumed, suggestively due to the sense of illegality and marginality that drug trafficking evokes. On the other side of the Atlantic, an informant reflects retrospectively also on the connotation of speaking of "porro" and how a cultural change has taken place (see Table 5.6).

The last remark of the informant aligns with the findings of international organizations. According to the United Nations, there was an increase of 23% in the number of people who used drugs between 2011 and 2021. Out of 296 million of users worldwide, 219 million consumed cannabis, 60 million opioids (of which 32 opiates), 36 million amphetamines, 22 million cocaine and 20 million "ecstasy" (UNODC, 2023). The alertness about the addiction dimension of psychotropic substances is reflected in the language used to address the topic, the vocabularies it activates and the references it mobilizes. As in the next example, in which, "droga" (drug) is used in the sense of pervasive and intense urge (see Table 5.7).

Describing an impending need as a "drug" is emblematic of the general perception held about addiction. The mere comparison of reading a newspaper, or spending time on other mundane unexciting activities with an addictive psychotropic seems out of proportion. Yet, it also illustrates to which degree the word "droga" (drug) is part of the biosocial repertoire of

Table 5.6 PRESEEA Oral Corpus excerpt on attitudes towards drugs in Seville, Spain

SEVI_M23_065 PRESEEA Corpus 2009 Seville, Spain Informant: Male, 37 years old	
I: <tiempo = "14:40"/> ¿la parte negativa del barrio? es que / en realidad Triana ha mejorado mucho con los años	I: <time = "14:40"/> the negative part of the neighborhood? is that / actually Triana has improved a lot over the years
E: uhum	E: um
I: o sea<alargamiento/> mmm antes de terminar las partes de Triana había bastan<palabra_cortada/ > bueno / había delincuencia / bueno había<alargamiento/> sobre todo<alargamiento/> mucho hurto / y ya no lo hay / o sea lo hay / como en todos lados / pero es una cosa generalizada / no es una cosa especifica / y después pues esta zona también tenía muy mala fama de prostitución y ya no la tiene // vamos / estoy hablando de<alargamiento/ > hace<alargamiento/> veinte o<alargamiento/> treinta años / pero<alargamiento/ > <ruido = "chasquido_ boca"/> que<alargamiento/> o<alargamiento/> por ejemplo en los ochenta / que se daba bastante<alargamiento/> el tema de la marihuana y el porro como una cosa / ofensiva o<alargamiento/> discriminatoria	I: I mean <lengthening/> mmm before finishing the parts of Triana there was enough <word_cut/> well / there was crime / well there was <lengthening/> above all <lengthening/> a lot of theft / and there isn't any anymore / I mean there is / like everywhere / but it is a generalized thing / it is not a specific thing / and then this area also had a very bad reputation for prostitution and it no longer has that // come on / I am talking about<extension/ > long ago<elongation/> twenty or<elongation/> thirty years ago / but<elongation/> <noise = "mouth_ click"/> that<elongation/> or<elongation/> for example in the eighties / that happened quite a lot<elongation/> the issue of marijuana and joint as a discriminatory thing / offensive or<extension/>
E: uhum	E: um
I: y ahora mismo pues como todos se están normalizando se están uniformando / pues lo ves de una forma normal no una fo<palabra_ cortada/> una cuestión de actitud / pero <vacilación/> o de elección / pero no <vacilación/> no lo veo mal / vamos	I: and right now, well, like everyone, they are normalizing, they are standardizing / so you see it in a normal way not a ma<word_cut/> a matter of attitude / but <hesitation/ > or of choice / but no <hesitation/> I don't see it as bad / come on

Table 5.7 PRESEEA Oral Corpus excerpt on the perception of the word “droga”

MADR_M21_024.MP3 PRESEEA Corpus 2002 Madrid, Spain Informant: Female, 42 years old	
E: háblame un poco de tu familia <simultáneo> a ver </simultáneo>	E: tell me a little about your family <simultaneous> let’s see </simultaneous>
I: <simultáneo> ¡ay! </simultáneo>	I: <simultaneous> oh! </simultaneous>
E: descríbeme un poco tu marido físicamente	E: describe your husband to me physically
I: <tiempo = “15:49”/> mi <vacilación/> / mi marido físicamente // bueno si le <vacilación/> / si le vieses / mmm dirías <cita> ¡esta mujer qué <entre_risas> exagerada es </entre_risas> </cita> / porque él tiene un aspecto / muy pacífico / que lo es / muy relajado / es un buen conversador es mmm / tranquilo / mm pero luego tiene / esto otro ¿no? / que le encanta sentarse en el sofá a leer su periódico / el periódico es como si fuera una droga yo ya le he deno <palabra_cortada/> <vacilación/ > denominado como que / es / mmm / ts una droga // eh l<alargamiento/>a televisión / eeh el dormir el descansar / entonces es una lucha // constante // porque esto como que lo apoya por la parte // de su familia ¿no? / ¿sabes? // y es muy duro / <simultáneo> es muy duro </simultáneo>	I:<time = "15:49"/> my <hesitation/> / my husband physically // well if you <hesitation/> / if you saw him / mmm you would say <quote> this woman is so <in_laughter> exaggerated </in_laughter> </quote> / because he has a very peaceful appearance / which he is / very relaxed / he’s a good conversationalist he’s mmm / calm / mm but then he has / this other side, you know? / that he loves to sit on the sofa and read his newspaper / the newspaper is like a drug I’ve already deno <word_cut/> <hesitation/> called it like / it’s / mmm / ts a drug // uh t<lengthening/>he television / eeh sleeping resting / so it’s a constant struggle // because it’s like he gets this support from // his family, you know? / you know? // and it’s very hard / <simultaneous> it’s very hard </simultaneous>
E: <simultáneo> ¿y<alargamiento/> físicamente </simultáneo> cómo es?	E: <simultaneous> and<lengthening/> physically </simultaneous> what does he look like

speakers of Spanish. Those in charge of educating the public know, nevertheless, that “droga” is just an umbrella term that escapes the complexities of the lexicon deployed to refer to recreational drugs.

Given the nature of substances with narcotic, anxiety- relieving and stimulant effects, it is logic that speakers create their language to refer to them, nonetheless because they move across clandestine circuits of illegal procurement, unregulated production and banned circulation. To create a glossary of street names of drugs, a rehabilitation center based in Barcelona opted to compare the jargon compiled by the sanitary authorities in Valencia in 1993 and present-day argot. Surprisingly, "no hay nada, bueno, casi nada" [*there is nothing, well, almost nothing*] (Borasteros, 2017, n.p.) that a consumer of these substances would not recognize, even if it has been three decades since the brief dictionary was put together.[21] Meanwhile, in the United States, in a message targeting Spanish-speaking parents on the dangers of medicines often abused by young people, the Massachusetts Department of Public Health (DPH) (2023) put together a list of the street names of medicines potentially used by Latino youth to designate certain brands of opioids, depressants, and stimulants. As the notice reads, "Como padre, es importante que reconozca los nombres callejeros de estos medicamentos" [*As a parent, it is important that you recognize the street names of these medicines*] (DPH, 2023). As Table 5.8 shows, the broad range of linguistic production derived from the consumption of substances produced, circulated, and traded without following any scientific standardization protocols. Like the Icelandic slang terms, the words in the list, I suggest, are linguistic pharmacy by-products.

Slang used to refer to drugs, including ab- and misused pharmaceuticals, illustrates the need for speakers to create a lexicon that adapts to their needs in response to input from science and (illegal) commerce. A non-conventional lexicon that replaces terms coined within prescriptive frameworks, such as scientific terminology. The logistic operation of designing a term as a referent that can become a standardized form is displaced by informal, ad-hoc, serendipitous, creative, onomatopoeic-prone and randomized coinage methods. Secrecy, familiarity, and stigma materialize in words that dislocate registers linked to social status or social hierarchies. Technical or official terms are not attractive. Similar to drug-related slang in Icelandic, in Spanish, anglicisms inform or serve as a basis for a wide number of adopted jargons used to refer to illegal substances or misused pharmaceuticals. For Rodríguez González (1994), the main motivation is the cryptic nature of expressions used within "sociolectos marginals" [marginal sociolects] (p. 210). This somehow presupposes, though, that those consuming psychoactive substances share a series of commonalities that extend beyond the urge to use a particular drug or pharmaceutical. For example, similar educational background, related occupation, generational affiliation, and class. One element that is clear in the context of the language of pharmacy is the roles of networks of dissemination, either of knowledge or information. As several examples in the oral and written corpora consulted for this book indicate, media outlets play a crucial role in the mobilization of lexicon. In the case of so-called

Table 5.8 Street names of drugs in Massachusetts, US (2023) and Valencia, Spain (1993)[a]

Massachusetts, USA (2023)		*Valencia (1993) and Barcelona (2023)*	
Official name in Spanish and scientific name in English	*Street name*	*Common name in Spanish*	*Street name*
codeína	captain cody, lean, schoolboy, sizzurp, loads, purple drank, loads, doors & fours, pancakes and syrup	**alucinógenos** (hallucinogenic)	tripis, volcanes, micropuntos, ácidos, secantes, lsd.
Fentanilo (actiq, duragesic, sublimaze)	apache, china girl, china white, dance fever, friend, goodfella, jackpot, murder 8, tango and cash, tnt	**anfetaminas** (anfetamines)	anfetas, speed, ruedas, bennies
hidrocona o dihidrocodeinona (dilaudid)	vike, watson -387	**barbitúricos** (barbituric)	barbs, caramelos, goodballs, peanuts
oxicodona (oxycontin, percodan, percocet)	o.c., oxycet, oxycotton, oxy, hillbilly heroin, percs	**benzodiazepinas** (benzodiazepine)	downs, tranks, roches (rohipnol), jaimitos (rohipnol)
metadona (dolophine, methadose)	amidone, fizzies with mdma; chocolate chip cookies	**cannabis**	chocolate, grifa, goma, mierda, tate, weed, yenga
morfina (duramorph, roxanol)	m, miss emma, monkey, white stuff	**cocaína** (cocaine)	farlopa, nieve, perico, yeyo
oximorfona (opana)	biscuits, blue heaven, mrs. o, o bomb, octagons, stop signs	**heroína** (heroine)	brown sugar, caballo, horse, jamaro, jaco, potro, white lady

[a] According to Massachusetts Department of Public Health (2023) and Direcció General d'Atenció i Prevenció de la Drogodependència (1993), Ramsés Gallego (2017) and Manuel Mas-Bagà (CAT Barcelona) (2023).

substandard forms, such as cant and slang words, television series, films, social media posts, online forums, and many other venues have become important nodes for nonformal linguistic input. Following my analogy that language resembles a logistic operation, I suggest that media outlets are important networks in the circulation of ideas about pharmacy in the 21st century. They come at the end of the line, after epistemic containers (knowledge) and epistemic labels (terms) have materialized in objects and practices. Reviewing briefly the case of Ozempic, and Rapamycin, in the next section, I address the role of mass communication in the linguistic adoption of pharmacy-related concepts.

5.3 Pharmaceutical Aural and Oral Side Effects

By shaping daily lives in different ways, the impact of pharmaceuticals from a biocultural and, as I suggest in this book, linguistic perspective is more than one might anticipate. Medicines can be taken as weight-management pharmacological options (e.g., semaglutide) and as possible antiaging medicines (e.g., Rapamycin). What is interesting about these two substances, and rather complex, names, is that their mediatic attention has *catapulted* them into the therapeutic, and hence linguistic, imaginary of millions around the world. Almost magically, pharmacy seems to offer products that could potentially help us to achieve the desired waistline, slow down metabolic aging and restore any physical marks of aging on the surface of our skin. But since medicines are the materialization of epistemic and commercial innovations primarily designed to potentially manage and/or cure diseases, it is not surprising that they eventually have implications beyond the context of health.

Additions in pharmacological knowledge logically require new words to label them, discuss about them, register them across systems, inform potential users about their existence and report on their efficacy, effectiveness, and adverse side effects. Perhaps due to the massive attention paid to body ideals in the 21st century, it seems normal to expect that a line of branded pharmaceuticals designed to medicalize weight loss has gained media attention. Medicines to control the feeling of hunger are popular. Social media platforms have played a role in spreading the word, for instance, on the existence of Ozempic (alternatively commercialized as Wegovy and Rybelsus). Produced by Novo Nordisk and approved in 2021 for "chronic weight management" (FDA, 2021), Ozempic has the potential to become a biocultural phenomenon, although not necessarily for the best reasons. It is too soon to establish how a chemical counterpart of glucagon-like peptide-1 (GLP-1), produced by humans, can be used to suppress hunger, or provide a feeling of satiety on a long-term basis. Semaglutide, with different names, is commercialized as, either a diabetes- or an obesity-management preparation. Physiologically, adverse side effects of taking them include nausea, constipation, and other gastrointestinal issues, besides an increased risk of developing kidney failure. Cosmetically, according to media reports, semaglutide is linked to a type of facial aging, "now known colloquially as "Ozempic face" (Suran, 2023, p. 167). According to expert Fatima Stanford, the "term is a misnomer as it presumes that facial aging is specific to semaglutide", whereas instead, "skin sagging is due to the loss of fat in the face, which is associated with any method of weight loss" (p. 167). After the news spread on the possible side effect of using semaglutide, almost instantly, *Ic.* "ozempic andlit" and *Sp.* "cara ozempic" (Ozempic face) emerged as colloquial terms across social media posts. By 2023, the language of pharmacy in Icelandic and Spanish had colloquially adopted a new term, a linguistic by-product of a pharmacological innovation.[22] The coinage of the portmanteaus in both languages simultaneously denotes the impact of media, commerce, and science in today's world. And while the plasticity of language to expand its lexicon in sync with novelties is positive. There might be a dark side to this as well. From another vantage point, "Ozempic face" can be regarded as an example of how language amalgamates pharmacy and identity, resulting in communicational mechanisms for cyberbullying, body shaming and name-calling across digital spaces.[23]

Pharmacy as the vector for the applications of scientific discoveries means that the language about medicines, somehow summarizes in a word, large bodies of knowledge and years of research, as well as epistemic transfers between knowledge systems and medicinal therapeutic traditions. Semaglutide was originally developed from an attempt to mimic the hormone used by Gila monster lizards (*Heloderma suspectum*) to suppress their appetite and control their sugar levels for extended periods of time (Williams, 2020). Today, years of trial and error and thousands of pieces of data on the effect of semaglutide on diabetes patients are discursively reduced to the effect of an agonist of a receptor to accelerate weight management. Or this is, at least, the most salient aspect of Ozempic/Wegovy across the media landscape. Similarly, Rapamycin, a medicine often used in circumventing and preventing transplant rejection,[24] has become known for its alleged antiaging properties in mammals. Specifically for slowing age-associated decline.

The history of Rapamycin (trade names: Sirolimus and Rapamune) is equally interesting, as it is produced by a species of filamentous bacteria (*Streptomyces hygroscopicus*) first found by a Canadian team exploring Rapa Nui (Easter Island) in 1964.[25] Prior to any intense contact with the rest of the world, before the opening of a major airport at the island, a team of scientists organized an expedition to conduct research on microbes and carcinogens found in "modern-day" products. While there, they noticed that, despite the high number of horses, the spread of tetanus bacteria (*Clostridium tetani*) was surprisingly low. After delving into the matter, Georges Nógrády and colleagues found one of the most powerful antifungals ever discovered, and labeled it "Rapamycin" in 1975. A combined term, mixing the reference to Rapa Nui with the Greek affix "-mycin", used commonly to refer to antibiotic compounds derived from fungi. Rapamycin has prompted other important developments, such as the mTOR inhibitors used for cancer, even if "few oncologists know that the familiar acronym 'mtor' refers to Rapamycin" and "fewer still know of its embedded reference to Easter Island" (Duffin, 2019, p. 350). The reflection on the origin of Rapamycin resonates with the origin of the name aspirin discussed in Chapter 1, the phonetic notation used for Indigenous medicinal plants discussed in Chapter 2 and the infixes discussed in Chapter 3. What is in a name and which references does the language of pharmacy mobilize? Rapa Nui (Easter Island) is probably not known to those benefiting from the fact that Rapamycin "was the beginning of an entirely new class of treatments" and the topic of "more than 40,000 articles" that explore many of the medicine's derivatives "applied to malignancies and immune therapies around the world" (p. 351). And while some indeed "tout it as a potential source of longevity, an elixir of eternal youth" (p. 351), without enough evidence (as yet), the word "Rapamycin" might change its meaning in the therapeutic landscape, and health-sphere, of the decades to come.

From a linguistic perspective, it demonstrates the logistic dimension of scientific terminology encapsulated in the amalgamation of a Polynesian and an Indo- European lexeme to create a meaningful descriptor. Yet, Rapamycin is not a fungus from Rapa, or it is, but not in the strict sense of the word. The baptism of the medicine recalls Kripke's (1990) definition of the baptism of a name (based on the transcript of the lectures he gave at Princeton University in 1970):

> An initial "baptism" takes place. Here the object may be named by ostension, or the reference of the name may be fixed by a description. When the name is "passed from link to link", the receiver of the name must, I think, intend when he learns it to use it with the same reference as the man from whom he heard it.
>
> (Kripke, 1990, p. 96)[26]

For it is in names that pharmacy substantiates the accumulation of knowledge. Pharmaceuticals become part of culture through linguistic units. Language facilitates the process by which pharmaceuticals become part of the therapeutic imaginary of communities. As Åsberg and Johnson (2016) observe in the case of Viagra, "While subjectivity is neither universal nor ahistorical, it is the product of structuring social relations, shaped by materialities and physical embodiment as much as by social phantasy" (Åsberg and Johnson, 2016, p. 89). Any form of information and promotion on innovations in pharmacy needs to be embedded in the context of expectations at an individual level and collective imagination. Yet, pharmaceuticals are not available to everyone. Brand names are proxy for safety, efficacy, and trust; knowing what your medications contain and where they come from can literally be a matter of life and death in many cases. The reality though is that the black market of pharmaceuticals in the world amounts to USD 4.4. billion.[27] Just as drugs developed by large pharmaceutical corporations and sold in certified dispensaries and pharmacies are part of the health ecosystems across the planet, so are counterfeit medicines. Also in that context, language plays a crucial role, as the names of unregistered pharmaceuticals equally rely on linguistic tactics to mobilize substandard and potentially dangerous preparations. Finding a convincing name, even from within the unauthorized framework in which this happens, illustrates nevertheless very well to which extent the language of pharmacy resembles a logistic operation.

Notes

1 Notes
The need to seek information prevails. According to Google trends, whereas the top searchers of the word "Viagra" reside in Italy, United Kingdom, and Germany, within the Spanish-speaking world, the medicine is most searched in Chile, Bolivia, and Mexico (Google Trends, 2023).

2 See for instance the discussion in Johnson, Sjögren and Åsberg (2016).

3 In addition to sildenafil, similar medications and brands include *vardenafil* (Levitra, Staxyn), *tadalafil* (Cialis), and *avanafil* (Stendra).

4 Furunbao is "un producto que se oferta al público como poseedor de supuestas propiedades como estimulante sexual. Según la información disponible, este producto se adquiere a través de Internet o por teléfono, y se desconocen los datos de las empresas que realizan su distribución en España" [*a product marketed to the public as having supposed properties as a sexual stimulant. According to the available information, this product is purchased online or by phone, and the details of the companies distributing it in Spain are unknown*]. Instead of natural ingredients, an analysis carried out by the Infarmed (Portugal) in 2014 has confirmed it contains tadalfile, without mentioning it on the labels, making it dangerous for patients with heart conditions and/or cardiovascular problems (AEMPS, 2015, p. 1).

5 See the study conducted by Ho *et al.* (2022), where the team demonstrates that, despite tight regulations, surveillance and lengthy approval protocols, in the 21st century, "(s)ubstandard and falsified medicines (SF) is a growing threat to the healthcare system" (p. 1) an that this also applies to herbal medicines.

6 See Li and Liu (2017) on traditional Chinese medical therapies for erectile dysfunction.

7 Faure (2014). Furthermore, on this, Gangwal and Gangwal (2011) add: "The well-selected name Viagra comes from prefix 'vi' connoting vigor and vitality, while the tail portion comes from Niagara, conjuring images of the power and fury of Niagara Falls. In this way perhaps the perfect name for a drug medically recommended for an erectile dysfunction. Levitra, the Viagra competitor, comes from elevate and Le indicates masculinity in French, and vitra can refer to vitality. In India, a version of Viagra is called Silagra, from its generic name, sildenafil" (p. 4).

8 See Johnson, Sjögren and Åsberg (2016) on the dynamics surrounding Viagra as a medication across the pharmaceutic-cultural landscape in Sweden. On one hand, from a nation-manhood nexus, and on the other hand, from an idealized representation of prosperity, modernity, access to quality healthcare and freedom. https://library.oapen.org/bitstream/handle/20.500.12657/24235/1005896.pdf?sequence=1.

9 On this point, Simon-Maeda (2008) observes: "The sexual pharmacology industry depends on advertising strategies which highlight prevalent notions of (hetero)sexual relationships wherein effective performance of the 'power tool' is conflated with ideas of 'authentic' masculinity" (p. 25).

10 Whereas the grammatical gender used for Vardenafil, another medicine used for erectile dysfunction is consistently masculine, for instance: "el mercado hay otras dos moléculas para la disfunción eréctil, el tadalafil y el vardenafil, que el año pasado vendieron 534. 000 y 44. 000 unidades" [*the market there are two other molecules for erectile dysfunction, tadalafil and vardenafil, which last year sold 534,000 and 44,000 units*] (Argentina 2018).

11 Makelawen pharmacies, Aliaga (2007) notes, do "not look like an immediate threat to Chile's retail pharmacies, which are dominated by three major chains, but it is gaining followers" (n.p.).

12 For many, the major milestones include the Single Convention on Narcotic Drugs (1961), the Convention on Psychotropic Substances (1971) and the United Nations Convention Against Illicit Traffic in Narcotic Drugs and Psychotropic Substances (1988). Although these were preceded by other forms of prohibition of recreational and mind-altering substances in Western countries, such as the British Pharmacy Act of 1868 and the alcohol prohibition in some countries, as well as the ban on consumption of opium and morphine in the first two decades of the 20th century.

13 See Sanchez Petrement (2023) on the cultural, historical, discursive and political context of a potential psychedelic renaissance in the 21st century.

14 Kvaran (2004b) has traced the first discussion on slang in Icelandic to Guðmundur Finnbogason's book *Íslendingar*, published in 1933. Specifically, to this paragraph on page 152: "Það mun vera orðgnótt, frjósemi og formfestu íslenzkunnar að þakka, að í henni hefir jafnan kveðið lítið að slanguryrðum (slang). Eins og prófessor Jespersen sýnir fram á, er slangurmálið sprottið af því, að menn verða leiðir á hversdagslegum orðum og gera sér því leik að því að bregða út af málvenjunni, með því að mynda ný orð og orðtök, er stinga í stúf við daglegt mál, stundum jafnframt til þess að vera ekki eins og aðrir..." [*It can be attributed to the richness, fertility, and structural consistency of the*

Icelandic language that it has generally been marked by a scarcity of slang terms. As Professor Jespersen points out, slang arises from people growing tired of ordinary words and therefore making a game of deviating from linguistic norms by creating new words and expressions that stand out against everyday speech, sometimes also as a means of distinguishing themselves from others...]. See Óskarsson (2022) on the history of the term "slangur" and similar concepts in Icelandic.

15 In a book review of the dictionary, Liberman (1988) remarks: "In comparison with the elaborate dictionaries of English slang and jargons, the Icelandic dictionary may strike one as a humble beginning. But it may well be that slang in general and specific cant words play a more modest role in Icelandic than in English. In any case, this is a useful and interesting reference book whose appearance should be welcomed" (p. 89).

16 In the case of English, Mattiello (2008) has shown that dialectal forms, as well as "bad language", cant and jargon often overlap with lexical formations, perceived as slang.

17 On Icelandic slang, cant and jargon, see also Liberman (2019), Héðinsdóttir (2014), Stong (1987) and Jones (1964).

18 In terms of loan words in Icelandic, Haselow (2012) asserts that their number is "kept to a minimum, by which the highly consociated character of the Icelandic lexicon could be preserved", but obviously, this does not seem to apply to slang.

19 Presumably related to "stoned" in English, which was originally "associated with the stupefaction induced by drugs but has developed towards a general description of drug- or alcohol-induced states" (Holme, 2012, p. 19).

20 According to the Online Etymology Dictionary (2023), "dope" can be traced to the 19th-century practice of "smoking semi-liquid opium preparation". In the Icelandic historic corpus, it appears 44 times.

21 In the format of an informal test, Singer Ramsés Gallego was asked to assess whether the terms were still commonly used for an interview and confirmed most words were still used in 2017 (Borasteros, 2017).

22 See, for instance, Arnþórsdóttir (2023), Miller (2023) and Sánchez (2023).

23 As many of the posts on social media, search engines and internet publications that include the term "Ozempic face" indicate.

24 Although, rapamycin "is currently not part of standard regimes for immunosuppression in renal transplantation, mainly due to its side effects, which include hyperlipidemia, pneumonitis and impaired wound healing" (Hobby *et al*, 2022, p, 1971).

25 Sirolimus is the name of a new compound identified in 1972 and "derived from *Streptomyces hygroscopicus* in the soil of Vai Atari, which is at the eastern end of the island next to the now quiescent volcano Rano Kao" (Kahan, 2003, p. 623).

26 Based on lectures given between 1970 and 1973, Kripke offers an alternative: "There is also an appeal to an initial baptism which is explained in terms either of fixing a reference by a description, or ostension (if ostension is not to

be subsumed under the other category). (Perhaps there are other possibilities for initial baptisms.)" (p. 97).

27 This is an Interpol's (2022) estimation. To illustrate the magnitude of the problem, only during a week-long operation, globally, "law enforcement made more than 7,800 seizures of illicit and misbranded medicines and healthcare products, totalling more than 3 million individual units" (n.p.)

Conclusion

Pharmacy, as a science and body of knowledge focuses on the preparation and administration of substances that have biological effects on living organisms, and inherently requires a specialized language. This intersection of language and medicine is crucial because pharmacy permeates the lives of patients, caregivers, and communities. Thousands of medicines are developed, produced, and consumed daily, with patients spending trillions of dollars on pharmaceuticals. This widespread presence has linguistic implications. Despite the complexity of pharmacy-related concepts and medicine names, they regularly appear in everyday conversations, becoming part of our therapeutic imaginaries and linguistic repertoires. The specialized vocabulary used when discussing pharmaceuticals and remedies reflects their role in our lived experiences. Almost everyone has taken a pill, received an injection, purchased an over-the-counter tablet, or used a plant-, mineral-, or animal-based substance for health management. Pharmacy, not just pharmaceuticals, shares a long history with humanity, embedding pharmacological knowledge in local medicinal practices worldwide. It's inevitable that we talk about the things we use to cure ourselves, especially when it becomes a matter of life and death.

Throughout the book, I have suggested that the language of pharmacy in Icelandic and Spanish could be viewed as a management process embedded in the organization and implementation of a complex operation: Communicating and registering the knowledge system of its speakers. From this perspective, language functions as a logistic undertaking. Like a parcel finding its way to its receiver, linguistic units travel across a managed flow from points of origin to points of reception, supported by the adaptability of natural languages. As I demonstrate in this book, the language of pharmacy is planned and managed similarly to a logistic operation.

DOI: 10.4324/9781003601654-7

It is crucial to remember that pharmacy and pharmaceuticals are not synonymous. Health is a vital element in the relationship between societies and knowledge. The prevention, diagnosis, treatment, and eradication of diseases are central to the survival strategies of every society worldwide. Consequently, all languages have developed ways to communicate, describe, and discuss health-related practices and objects. In the Western context, medicinal, pharmaceutical, and therapeutic knowledge is linguistically expressed in harmony with the advances, innovations, and discoveries driven by empirical research. This alignment explains the exponential growth of biomedical language in the 20th and 21st centuries, a trend likely to continue.

For many, this has led to increased exposure to unfamiliar terms, which have gradually become part of their cultural repertoire. In contemporary Western societies, health now encompasses more than merely the absence of illness. Biomedical terms are commonly found on the labels of food items, cosmetics, and toiletries. Language plays a crucial role in moderating how medicinal efficacy is understood, perceived, and marketed. The omnipresence of health and its commercial potential has significant implications for the general population. Culture is influenced by science, and scientific language undergoes cultural adoption and adaptation. Like any logistic operation, biomedical language requires stages of procurement, maintenance, and transportation before reaching its various final destinations. If the language of pharmacy is viewed as a logistic operation, it can be compared to the use of containers, labels, and networks. Throughout this book, I have introduced the concepts of epistemic containers, epistemic labels, and epistemic networks to illustrate this process. Epistemic containers refer to the contexts that encapsulate the organization of scientific language and traditional knowledge reporting. These include descriptions, discussions, controversies, interactions, and negotiations captured by language whenever there are changes or additions in a knowledge system. Containers, such as scientific journals, are crucial for understanding the historical development of the language of pharmacy. Once consensus is achieved, epistemic labels emerge. These labels are the standardized terms and references institutionalized by language, used to structure and communicate new knowledge and innovations. In pharmacy, commerce plays a significant role in creating these labels through brand names designed to attract attention and establish the identity of pharmaceuticals (e.g., aspirin). Subsequently, these labels enter organized networks, which are complex systems that facilitate the flow of knowledge from its origin to linguistic communities. For instance, while "vitamin" serves as an epistemic label, terms like "trophone", "sitacoid", "advitants", and "accessory food factors", discussed historically as alternative scientific terms for vitamins (Ramsden, 1918; Brown, 1920; Maltz, 1992), represent pieces of an epistemic container. A modern English dictionary entry for "vitamin" exemplifies an epistemic network, showing how labels are integrated into broader informational systems.

Science relies on the presentability and representability of knowledge to enable meaningful exchanges that drive its progression. In this regard, it is similar to any other form of knowledge. It requires a facilitating mechanism for transmission, typically through some form of communication. This is where language, knowledge, and science intersect. Elements in pharmacy can originate from trial-and-error episodes, whether they occur in a laboratory or during a healing intervention in a non-urbanized setting. These "epistemic episodes" are sites where object-dependent knowledge is produced. Whether in the form of a formula or a plant-based concoction, a biological agent is essential for pharmacy to exist.

Since this necessity spans all therapeutic treatments based on knowledge, including those in non-scientific contexts (by Western definitions of "science"), all languages naturally have words to express concepts such as "medicine", "drug", "healing object", "healing precursor", "potion", and "remedy". The language of medicine has evolved to accommodate the vocabularies needed to discuss healing approaches since ancient times. In the context of Western biomedicine, which serves as the starting point of my argument, history has documented the onset, development, and evolution of pharmacy as a discipline and a field of life-saving knowledge production. Simultaneously, pharmaceuticals, as end-products, are central to commercial enterprises and networks, often linked to profit-driven medicalization initiatives and the significant economic gains generated by the commercialization of medicines. Pharmacy and commerce are deeply intertwined. Developing and producing medicines require substantial capital, and even in state-run healthcare systems, the costs must be covered by someone along the line.

From the 17th century, European vernacular languages evolved under the influence of scientific inquiry and debate, leading to a new emphasis on logical, depersonalized objectivity (Halliday and Butt, 2019). This period also saw European colonization, which pressured languages to reflect material and epistemic exchanges. The natural sciences drove curiosity and the acquisition of goods, integrating objects, practices, concepts, and ideas from distant regions into European culture, politics, and languages. Additions such as materia medica from the Middle East, Far East, and the Americas significantly impacted early European pharmacy, with each "curiosity" acquiring a name. For example, while André Thevet may have introduced tobacco to Europe, it was Jean Nicot who popularized it as a medicine in the 16th-century Medici court, leading to the term "nicotine" rather than "thevetine". The names used in Brazil at the time, like "petun", did not survive in modern pharmaceutical language. This demonstrates how logistic operations determine the linguistic solidification of science and knowledge. Today, scholars, scientists, and commercial influencers shape the language used in pharmacy. Scientific publications remain crucial for disseminating terminology and negotiating linguistic adaptations of scientific innovations. Trade names, often overlooked in natural language studies, are significant in pharmacy-related contexts. For instance, the metaphor "viagra de la literatura" (literary viagra) is effective due to Viagra's established role in the medicinal repertoire, illustrating the pharmaceutical language's reach beyond health contexts.

Pharmacy's influence on language extends to morphology and phonetics. Howard Ansel (1965) noted that pharmaceutical manufacturers often create distinctive and memorable names, which can confuse physicians and pharmacists, forcing them to become "amateur etymologists" to understand the pharmaceuticals they prescribe and sell. If linguistics involves the systematic and comprehensive analysis of language, it might also be valuable to explore how certain types of language emerge from logistic processes. As discussed throughout this book, the language of pharmacy exemplifies the creation and maintenance of a linguistic system through a series of stages, all contributing to the completion of a specific task. For instance, as I discuss in Chapter 1, the term *ibuprofen* went through various stages before becoming part of the therapeutic vocabulary of someone suffering from a headache. This process involved initial scientific knowledge production, followed by discussions about the term (involving both science and commerce), which then led to the word's circulation through commerce and marketing, and its eventual adoption into modern English. This term has since been integrated into hundreds of other languages, including Icelandic and Spanish, as demonstrated by the corpus entries.

Linguistic planning strategies at both micro and macro levels solidified alongside the establishment of various fields in Iceland during the 18th, 19th, and early 20th centuries. This process was closely tied to the introduction of cutting-edge scientific discoveries and innovations from Europe and North America, often facilitated by print media practices such as translation. Knowledge brokers translating foundational academic volumes and textbooks played a crucial role in introducing Icelandic readers to an unprecedented number of new scientific terms, concepts, technical terms, loanwords, and neologisms, often densely packed within single paragraphs. The term "efnafræði" became established as the consistent designation for the field of chemistry through a process like the one by which many words were incorporated into Icelandic in the 19th and early 20th centuries. Influential figures, often educated in Denmark and residing in Copenhagen for years, actively contributed to the expansion and adoption of terminology reflecting the latest scientific innovations.

The impact of these intellectuals on the creation of Icelandic neologisms in response to chemical advancements in pharmacy mirrors the significance of works on pharmacognosy authored by Mexican doctors. In both contexts, language facilitated the moderation, articulation, and dissemination of knowledge.

Entries in Nahuatl, Maya, and Otomi, among other Indigenous languages, found in 19th and early 20th-century Mexican pharmacopoeias, highlight the scientific, scholarly, colonial, political, and commercial processes involved. These entries also demonstrate the critical role of linguistic planning in the transfer, co-production, and co-dissemination of medicinal knowledge. To this day, in many societies, the awareness of pharmacological qualities found in nature constitutes a compendium of information that has been preserved. This knowledge transfer, likely dependent on trial and error as well as language, involves instructing and informing others about medicines with a certain level of descriptiveness and detail. Oral tradition continues to play a crucial role in the application of natural medicine in various parts of the world. In 19th-century Mexico, as in 18th and 19th-century Iceland, language was pivotal in understanding and communicating scientific knowledge. The dissemination of pharmacy-related knowledge was managed through language, confirming the similarities between science-driven linguistic approaches and task-oriented logistic operations. Icelanders needed to embrace and adapt to chemistry as a field, just as Mexicans needed to adopt and adapt pharmacological knowledge compilation protocols. In both cases, finding or designing the right words was essential.

Morphologically, much of the scientific lexicon is constructed from units carefully assembled to convey information at various levels. As I have shown through various examples, even a small morpheme, such as an infix, can provide insights into the history and epistemic systems from which it originates. The linguistic presence of quinine (cinchona), for instance, registers the Quechua ancestry of therapeutic practice and chronicles the journey of the bark through European medical cabinets and laboratories. The persistent presence of **-quin-** as an infix in contemporary chemistry serves as a reminder of the knowledge transfer from the Andean medical repertoire; an ancestral knowledge system. The use of quinaquina as a febrifuge obtained from the Cinchona genus trees throughout the Andean region preceded the experimentation, trial and error, and observation protocols to which the bark was later subjected in European and North American laboratories. Although different in methodology, similar processes likely took place across the Andes before the bark received the name "bark of bark"/"holy bark"—"quinaquina". The movement of the bark through imperial networks included a linguistic component that has survived for centuries. The Quechua legacy of the molecular composition of quinine [(9R)-6-methoxy-8α-cinchonan-9-ol], in its various nominal forms (e.g., quinaquina, quina, kina), remains in modern-day chemical nomenclature. More importantly, it has retained its place in the knowledge system where it first emerged. Focusing specifically on **-quin-** as an infix, as seen in chloroquine, Atovaquone (quone > quinine + one), or Atoquinol, is not about deconstructing terminologies in the manner of Saussure or Derrida. Instead, it highlights how linguistic evidence can contextualize the creation of the language of pharmacy. In this sense, language serves as a marker of the epistemic identity of raw materials used for medicinal purposes. I have used the term epistemic identity to refer to the processes, sites, and actors involved in the knowledge production behind therapeutic and pharmacological innovations. Regardless of their origin, language plays a role in any strategy devised to pass knowledge on to others. Health-related vocabularies exist in all languages, confirming that no society is devoid of a medicinal repertoire. Applying knowledge to restore human health is a shared heritage of both biomedicine (Western science) and traditional epistemic systems, such as the one in which quinine first emerged to treat fevers. In many ways, the Quechua-derived infix **-quin-** in today's chemical nomenclature is historically, epistemically, linguistically, and culturally significant because it remains part of the modern therapeutic repertoire.

Language operates outside the biopolitical dimension of classification, at least in the sense of pharmacy exerting power over existing entities. Pharmacy has existed long before pharmaceuticals became the technological materializations, we understand them to be in the 21st century. Objects and subjects, including plants, hold a place in the ontology of societies aware of their existence. From a Heideggerian perspective, nature's potential to become medicinal anticipated any coordinated scientific project derived from knowledge accumulation rather than mere survival tools. It is thus logical to assume that the language of pharmacy also preceded modern technology and science.

In its most basic form, pharmacological preparations could have been communicated through a hand gesture, exclamation, or shout. Every language has a system to articulate information about remedial and curative preparations, relying on the distinction between substances. Linguistic distinctions induce epistemological categories in various formats, shapes, dimensions, associations, orders, or disarrays. Traditional pharmacological knowledge does not impose reformulated categories or erase existing entities. The same holds for pharmaceuticals derived from biomedical and chemical experimentation. In both cases, the names of medicines are designations and references to the outcomes of trial-and-error episodes, practices, and recorded emulation, variation, and adaptation.

On this point, the history of the discovery, synthesis, and classification of the thirteen known vitamins illustrates the intricate relationship between scientific language, commerce, and the dissemination of information through media. Pragmatically, this is evident in the use of specialized lexicon within different registers. In certain contexts, detailed terms such as "riboflavin" and "ascorbic acid" are used instead of the more common "vitamin B_2" and "vitamin C". More importantly, as scientific terminology is integrated into everyday language, it reveals the dynamics of how specialized terms evolve in non-scientific settings. For instance, the phrase "vitamin injection" (*Ic.* vítamínsprauta) is used to describe an incentive or stimulus. This continuous linguistic adaptation highlights how communities are influenced by scientific and technological advancements. In the context of pharmacy, language is meticulously managed from its initial creation (coinage) to its eventual adoption. As I have demonstrated in the preceding chapters, the commercialization of medicines is a major source of linguistic production. Terms for pharmaceuticals originate from highly scientific settings (e.g., the term "vitamin") as well as from marketing- driven circuits (e.g., Viagra). These processes involve multimillion- dollar investments in top- notch facilities, rigorous procedures, and expensive raw materials. Even more is at stake when considering the health of billions of people and the precarious situation of those who cannot access medicines due to their high costs. Pharmacy exemplifies the commercialization and materialization of knowledge, a double- edged sword and a relatively recent development. While medicines are essential for preserving and restoring health, many individuals are excluded due to high costs. The connection between medicines, commercial interests, and language is evident. Potential buyers, especially for over-the-counter products, need to know and understand the medicines they proactively seek. It is through a specific type of language that medicines are marketed and presented to consumers. For these consumers, a proper noun denoting an object with pharmacological properties becomes part of their linguistic repertoire, serving as a referent and acquiring semantic and pragmatic meanings.

The connection between language and pharmaceuticals is best exemplified by the strategies used to brand medicines. Although the relationship between health and pharmacology existed long before the concepts of commercialization and laissez- faire economics, pharmaceuticalization (the increased use and availability of medicines) is undeniably profitable. Selling medicines is a multimillion-dollar industry, and language is a key tool in advertising. Before patients can be prescribed medicines, doctors must have essential information about the pharmaceuticals on the market, including their names, medical uses, efficacy, and pharmacology. Language is crucial in medico-marketing because it relies on terminology that targets specialists. While language moderates science communication, semiotics plays a decisive role in disseminating information about pharmacy.

Pharmacy is linguistically articulated through names that highlight scientific perception while downplaying the commercialization of medicines. More importantly, the products of pharmacy become artifacts that circulate through society, prompting reactions on legal, cultural, and linguistic levels. Language records and enables speakers to adapt to, adopt, and articulate the influence that medicines have in their bodies, lives, and communities. As I have shown, pharmacy generates linguistic by-products. This explains why the commercialization of medicines has had a linguistic impact on Icelandic and Spanish speakers at various levels and in unexpected ways. Surpassing the boundaries of prescriptive language, oral and written references to brand names, pharmaceutical, and scientific terminology illustrate how the legal, economic, cultural, and biosocial aspects of medicines—substances for the care and cure of the body—are adopted sociolinguistically. This applies, for instance, to the slang related to psychoactive substances and sexual pharmacy terminology. As the excerpts from the corpora and transcribed interviews show, the linguistic implications transcend scientific considerations and regulatory protocols. Even though the language of pharmacy is logistically managed, how speakers actually use it escapes any predetermined schematization. Health-restoring and health-preserving objects, whether derived from pharmacy or traditional knowledge (or any other form of medicine), encapsulate technoscientific innovation and epistemologies (of various sorts and various levels). They also serve as entry points into the realm of language as a method, system, and means of communication—one that I suggest is logistically managed, planned, and executed.

This book has examined the morphological and semantic-pragmatic shifts in language brought about by pharmacy and traditional medicinal knowledge, considering epistemic paradigms as linguistic factors, the optimization prompted by interactions between different knowledge systems, and the linguistic distinctiveness and specialization surrounding pharmacy-related objects in media, particularly in oral and written Icelandic and Spanish. My central argument has been that the language of pharmacy operates as a logistic task. As shown, creating linguistic conventions about medicines involves the planning and execution of steps such as transaction, addition, renewal, expansion, extraction, and especially the transportation and movement of epistemic containers and labels. In this context, containers refer to bodies of knowledge and discoveries about pharmacological processes, while labels are the terms (concepts, scientific terms, medicine names, brand names) that constitute the language of pharmacy. These often emerge from scientific circles and are coined to reflect procedures and contexts of knowledge production, and provide essential taxonomic information. Linguistic conventions are fundamental to pharmacy, a field that amalgamates science, knowledge, and commerce in complex ways. Using historical written sources, various media formats, and natural language oral corpora in Icelandic and Spanish, I have demonstrated that scientific input through the language of pharmacy significantly influences speech communities sociolinguistically. Additionally, I have provided plenty of examples from the corpora that show the ways in which the language of pharmacy has retained input from vernacular, traditional, and Indigenous knowledge systems.

The linguistic traces of communities whose medicinal legacies have contributed to creating medicines are, in some cases, still present in the language of pharmacy. The Amerindian origins of infixes like **-ca-**, **-cur-**, and **-quin-** in medicines used globally underscore this point. The chemical nomenclature applied to medicines highlights the connection between pharmacy and chemistry. The diverse origins of pharmacy-related lexicon further support the logistic aspect of language. This diversity also emphasizes that pharmacy and pharmaceuticals are not synonymous; labeling pharmacy as solely related to the pharmaceutical industry ignores the traditional medicine that has existed for millennia. Pharmacy lies at the nexus of science and commerce.

By examining the language of pharmacy and traditional medicinal knowledge, my aim has been to show the evolution of the language-science dyad in the 21st century Icelandic and Spanish. A historical and present-day analysis of the interaction between linguistic communities, science, technology, and knowledge transfers reveals that science drives language to function as a logistic process involving planning, management, and execution. Speakers react to these processes by including pharmacology-related lexicon in their language production, and adapting it. The careful management and planning of scientific language eventually result in the emergence of linguistic by-products, shaped by the culture of the speakers exposed to the incursion of science in their community. Ultimately, language is a window into our social life, where we become sick, need medicines, and talk about them in our terms.

Appendix

International Non-Proprietary Names (World Health Organization 2017):

> The INN system as it exists today was initiated in 1950 by a World Health Assembly resolution WHA3.11 and began operating in 1953, when the first list of International Nonproprietary Names for pharmaceutical substances was published. The cumulative list of INNs now stands at some 9300 names designated since that time, and this number is growing every year by some 160 new INNs. Since its inception, the aim of the INN system has been to provide health professionals with a unique and universally available designated name to identify each pharmaceutical substance. The existence of an international nomenclature for pharmaceutical substances, in the form of INNs, is important for the clear identification, sale prescription and dispensing of medicines to patients, and for communication and exchange of information among health professionals and scientists worldwide. As unique names, INNs have to be distinctive in sound and spelling, and should not be liable to confusion with other names in common use. To make INNs universally available they are formally placed by WHO in the public domain, hence their designation as "nonproprietary". They can be used without any restriction whatsoever to identify pharmaceutical substances. Another important feature of the INN system is that the names of pharmacologically-related substances demonstrate their relationship by using a common "stem". By the use of common stems the medical practitioner, the pharmacist, or anyone dealing with pharmaceutical products can recognize that the substance belongs to a group of substances having similar pharmacological activity.

References

19. Júní, 1924. '"Vitamin"-kenningin', 1 November.

Academia Médico-Quirúrgica, 1832. *Ensayo para la materia médica mexicana.* Puebla: Oficina del hospital de S. Pedro. Available at: https://digirepo.nlm.nih.gov/ext/mhl/61540040R/PDF/61540040R.pdf

Aceves Pastrana, P., 2007. 'La transformación de la tradición farmacéutica en México 1871–1930'. In *38th International Congress for the History of Pharmacy*, Sevilla, September 19–22, 2007.

Aftonbladet, 1954. 21 April.

Agencia Española de Medicamentos y Productos Sanitarios (AEMPS), 2015. 'Comercialización ilegal del producto furunbao'. 30 July.

Agrawal, M., 1993. 'International Pharmaceutical Marketing: An Historical Perspective'. Proceedings of the Conference on Historical Analysis and Research in Marketing, pp. 363–376.

Ahltorp, M., Skeppstedt, M., Kitajima, S., Henriksson, A., Rzepka, R. and Araki, K., 2016. 'Expansion of medical vocabularies using distributional semantics on Japanese patient blogs'. *Journal of Biomedical Semantics*, 7, pp. 1–18.

Akbari, F., 2020. *Iran's Language Planning Confronting English Abbreviations: Persian Terminology Planning.* Cham: Springer Nature.

Albury, N.J., 2016. 'National language policy theory: Exploring Spolsky's model in the case of Iceland'. *Language policy*, 15(4), pp. 355–372.

Alcaraz Varó, E., 2006. 'Aproximación al lenguaje de la Farmacia'. *Anales de la Real Academia Nacional de Farmacia*, 72(2), pp. 343–360.

Alfaro, R., 1866. 'Terapéutica del cihopatli o zoapatle'. *Gaceta Médica de México: Órgano de la Sociedad Médica de México*, II, pp. 47–48.

Aliaga, C., 2007. 'Chile rediscovers native Mapuche remedies'. Reuters, 26 March. Available at: www.reuters.com/article/us-chile-mapuche/chile-rediscovers-native-mapuche-remedies-idUSN2342701220070326

Al-Kasimi, A.M., 2019. 'The history of Arabic lexicography and terminology'. In: Kageura, K. and Temmerman, R. (eds.) *Handbook of Terminology*, Vol. 2, Amsterdam: John Benjamins, pp. 7–30.

Almarsdóttir, A.B. and Grímsson, A., 2000. 'Over-the-counter codeine use in Iceland: The impact of increased access'. *Scandinavian Journal of Public Health*, 28(4), pp. 270–274. Available at: www.jstor.org/stable/45137465

Alonso y de los Ruices de Fontecha, J. 1606. *Diez preuilegios para mugeres preñadas*. Alcala de Henares: Por Luys Martynez Grande.
Álvarez Jurado, M., 2021. 'En torno al prólogo y las notas de la traducción al español del Codex, Pharmacopée française de 1837'. *Çedille: Revista de Estudios Franceses*, 20, pp. 215–235.
Amador, E., 1899. *Apuntes para la flora zacatecana*. Guadalupe: Tipografia del Hospicio de Niños en Guadalupe.
American Pharmaceutical Association, 1894. *Proceedings of the American Pharmaceutical Association at the annual meeting*. Baltimore, MD: The Association.
Andvari, 1880. Copenhagen: S.L. Møller https://timarit.is/page/4323421?iabr=on#page/n1/mode/2up/search/berg [Access 7 December 2024].
Ansel, H.C., 1961 [1965]. 'Etymology of pharmaceutical trademarks'. *Journal of the American Pharmaceutical Association*, 5(3), pp.160–162. Available at: https://doi.org/10.1016/S0003-0465(15)33801-5
Apinis, P., Salaks, J. and Pozņaka, V., 2010. 'Medicīnas grāmatniecība Latvijā un Baltijā 1990–2010. gadā'. *Acta medico-historica Rigensia*, IX, pp.50–74.
Apple, R.D., 1996. *Vitamania: Vitamins in American Culture*. New Brunswick, NJ: Rutgers University Press.
Applequist, J. and Ball, J.G., 2018. 'An updated analysis of direct-to-consumer television advertisements for prescription drugs'. *The Annals of Family Medicine*, 16(3), pp. 211–216.
Arbetet, 1953. 15 September.
Árnason, K., 2002. 'Icelandic'. In: A. Deumert and W. Vandenbussche (eds.) *Germanic Standardizations: Past to Present*. Vol. 18. Amsterdam: John Benjamins, pp. 245–279.
Árnason, M., Sigmundsson, S. and Thorsson, Ö., 2019 [1982]. *Slangur*. Reykjavik: Forlagið.
Arnþórsdóttir, E., 2023. 'Notendur samfélagsmiðla hvattir til að nota sykursýkislyf í megrunarskyni'. *Heimildin*, 20 March. Available at: [https://heimildin.is/grein/17143/]
Åsberg, C. and Johnson, E., 2016. 'Viagra selfhood: Pharmaceutical advertising and the visual formations of Swedish masculinity'. In: Johnson, E., Sjögren, E. and Åsberg, C. (eds.) *Glocal Pharma: International Brands and the Imagination of Local Masculinity*. London and New York: Routledge, pp. 88–99.
Ásmundsson, P., 2014. 'Guðmundur Björnsson landlæknir —150 ára minning'. *Læknablaðið*, 100, pp. 628–630.
Aufegger, L., Yanar, C., Darzi, A. and Bicknell, C., 2012. 'The risk-value trade-off: Price and brand information impact consumers' intentions to purchase OTC drugs'. *Journal of Pharmaceutical Policy and Practice*, 14(1), pp.1–13.
Azorín Fernández, D. and Sánchez Manzanares, C., 2016. 'Propuesta de un diccionario de neologismos: cuestiones metodológicas y criterios lexicológicos'. In Cotelo García, R. (ed.): *Entre dos coordenadas: La perspectiva diacrónica y diatópica en los estudios léxicos del español*. San Millán de la Cogolla: Cilengua, pp.11–31.

Badie, M.C., Dell'Arciprete, A., Fontes, C., Hirsch, S. and Orlando, M.F., 2013. 'Uso, circulación y significación de los medicamentos en comunidades Pilagá, guaraní, MBYA-Guaraní, Tapiete y Toba'. *Revista Argentina de Salud Pública*, 4(17), pp. 13–23.

Ball, C. and Westhorpe, R., 2005. 'Muscle relaxants—intocostrin'. *Anaesthesia and Intensive Care*, 33(3), pp. 289. Available at: https://doi.org/10.1177/0310057X0503300302.

Balmaceda, D., 2020. 'La fascinante historia de Geniol y de la mítica publicidad de la cabeza con clavos'. Zonaelecasilda, 7 December. Available at: https://zonaelecasilda.wordpress.com/2020/12/07/la-fascinante-historia-de-geniol-y-de-la-mitica-publicidad-de-la-cabeza-con-clavos-por-daniel-balmaceda/.

Balmaceda, D. 2021. 'El curioso origen de Geniol y de la inolvidable publicidad de la cabeza con clavos'. *iProfesional.* 6 August. www.iprofesional.com/actualidad/344813-conoce-el-increible-origen-de-geniol-por-daniel-balmaceda [Access December 2024].

Bamyaci, M.E., 2021. 'Definition, importance and historical development of logistics from the beginning to the present'. In: Ünar, Ş. and Karagöz, S. (eds.) *Social and Humanities Science: Research, Theory*. Lyon: Livre de Lyon, pp. 243–255.

Banting, F.G., 1929. 'The history of insulin'. *Edinburgh Medical Journal*, 36(1), p.1.

Bautista Becerro, F., 2022. 'Análisis macrotextual, terminológico y traductológico de los textos farmacéuticos y estudio de campo de la traducción farmacéutica en España'. PhD Dissertation, University of Salamanca. Available at: https://gredos.usal.es/handle/10366/150741.

Baxerres, C. and Le Hesran, J.Y., 2011. 'Where do pharmaceuticals on the market originate? An analysis of the informal drug supply in Cotonou, Benin'. *Social Science & Medicine*, 73(8), pp. 1249–1256.

Becerro, F.B., 2020. 'Phraseological variations in medical-pharmaceutical terminology and its applications for English and German into Spanish translations'. *SciMedicine Journal*, 2(1), pp. 22–29.

Bellés Fortuño, B., 2016. 'Medical discourse in professional, academic and popular settings'. In Ordóñez-López, P. and Edo-Marzá, N. (eds.): *Medical Discourse in Professional, Academic and Popular Settings*. Bristol: Multilingual Matters, pp. 55–78.

Bello, R.M., 2016. 'La terminología química y los libros de enseñanza de la química durante la primera mitad del siglo XIX en España'. In: Garriga Escribano, C. and Pérez Pascual, J.I. (eds.) *Lengua de la ciencia e historiografía*. Universidade da Coruña: Servizo de Publicacións, pp. 193–206.

Bennett, B.C. and Balick, M.J., 2014. 'Does the name really matter? The importance of botanical nomenclature and plant taxonomy in biomedical research.' *Journal of Ethnopharmacology*, 152(3), pp. 387–392.

Berg, T., 1872. *Zur Kenntniss des in der Cetraria islandica Ach. vorkommenden Lichenins und jodblaüenden Stoffes*. Dorpat (Tartu): Druck von C. Mattiesen. Berlin–New York: Mouton de Gruyter.

Bertomeu Sánchez, J.R. and Muñoz Bello, R., 2010. 'Resistencias, novedades y negociaciones: la terminología química durante la primera mitad del siglo XIX en España'. *Dynamis*, 30, pp. 213–238.

Bertomeu Sánchez, J.R. and Muñoz Bello, R., 2012. 'La terminología química durante el siglo XIX: Retos, polémicas y transformaciones'. *Educación Química*, 23(3), pp. 405–410.

Bickel, M.H., 1988. 'The development of sulfonamides (1932–1938) as a focal point in the history of chemotherapy'. *Gesnerus*, 45(1), pp.67–86.

Bigaux, André (Ltd.), 1960 (ca.). Vioftalyl Leaflet.

Bigaux, André (Ltd.), 1984. *El Informador*, 16 August.

Bigaux, André (Ltd.), 1989. *El Informador*, 14 June.

Bjarnadóttir, K., 2012. 'The database of modern Icelandic inflection (Beygingarlýsing íslensks nútímamáls)'. In: De Pauw, G. et al. (eds.) Proceedings of the Workshop on Language Technology for Normalisation of Less-Resourced Languages. Istanbul: ELRA.

Bjarnadóttir, K., 2019. 'Orðaskrá úr orðabókarhandriti Jóns Ólafssonar úr Grunnavík'. Árnastofnun. Available at: https://arnastofnun.is/sites/default/files/2019-04/Or%C3%B0askr%C3%A1%20%C3%BAr%20or%C3%B0ab%C3%B3karhandriti%20J%C3%B3ns%20%C3%93lafssonar%20%C3%BAr%20Grunnav%C3%ADk.pdf.

Bjarnason (bónda), A., 1860. *Um íslenzkar drykkurtir, söfnun þeirra, geymslu, nytsemi, verkanir bg tilreiðslu*. Akureyri: Prentsmiðja Norður- og Austnrumdænisins hjá H. Helgasyni.

Bjerg Larsen, J., Mount, J.K., Kruse, P.R. and Vrangbæk, K., 2004. 'Dynamics of pharmacy regulation in Denmark, 1546–1932: A study of profession-state relations'. *Pharmacy in History*, 46(1), pp. 3–25.

Björnsson, G., 1912. 'Rangritunarheimska og framburðarforsmán'. *Skólablaðið*, 10, pp. 147–151.

Björnsson, G., 1913. 'Hljómbætur'. Skólablaðið, 12, pp. 181–186.

Björnsson, G., 1930. *Bráðabirgðareglugerð um nám í lyfjafræði og aðstoðarmannspróf*. Reykjavik: N.p.

Black, J., 1999. 'Claude Bernard on the action of curare'. *BMJ (Clinical research ed.)*, 319(7210), p. 622. Available at: https://doi.org/10.1136/bmj.319.7210.622.

Black, S.P., 2023. 'Ethics and language'. In: Manning, P. and Wilson, J. (eds.) *A New Companion to Linguistic Anthropology*. London: Wiley, pp. 299–314.

Blair, D.C., 1990. *Language and Representation in Information Retrieval*. Amsterdam: Elsevier.

Blane, G., 1780–81. *A Short Account of the Most Effectual Means of Preserving the Health of Seamen, Particularly in the Royal Navy*. London: N.P. Wellcome Collection, Digitalized copy. Available at: https://wellcomecollection.org [Accessed October 2023].

Blech, J., 2023. *Inventing Disease and Pushing Pills: Pharmaceutical Companies and the Medicalisation of Normal Life*. London: Routledge.

Bleichmar, D., 2015. 'The imperial visual archive: Images, evidence, and knowledge in the early modern Hispanic world'. *Colonial Latin American Review*, 24(2), pp. 236–266.

Bliss, M., 1993. 'The history of insulin'. *Diabetes Care*, 16(Supplement 3), pp. 4–7.

Boas, F., 1942. "Language and Culture." In *Studies in the History of Culture: The Disciplines of the Humanities,* edited by Percy Long, pp. 178–184. Menasha, WI: Banta.

Bokhour, B.G. and Kressin, N.R., 2015. 'What is in a name? How biomedical language may derail patient understanding of hypertension'. *Circulation: Cardiovascular Quality and Outcomes*, 8(4), pp. 452–454.

Borasteros, D., 2017. 'No seas un 'Goofy': el test del nuevo lenguaje de la droga'. *El Confidencial*, 4 July. Available at: www.elconfidencial.com/alma-corazon-vida/2017-07-04/el-coleta-nos-explica-cada-droga-menos-los-bennies-que-no-sabe-lo-que-son_1407962/ [Accessed October 2023].

Borell, M., 1978. 'Setting the standards for a new science: Edward Schaefer and endocrinology'. *Medical History*, 22(3), pp. 282–290.

Borgoño, C.A. and Zinman, B., 2012. 'Insulins: Past, present, and future'. *Endocrinology and Metabolism Clinics*, 41(1), pp. 1–24.

Botero Peláez, B., Acebedo Ferrer, O. and Montes Sáenz, J., 1968. 'El factor tubario en la esterilidad femenina'. *Revista Colombiana de Obstetricia y Ginecología*, 19(5), pp. 311–336.

Boumediene, S., 2016. La colonisation du savoir: Une histoire des plantes médicinales du "Nouveau Monde" (1492–1750). Vaulx-en-Velin: Les éditions des mondes à faire.

Bowen, W.H., 1938. 'The earliest treatise on tobacco: Jacques Gohory's "Instruction sur l'herbe Petum"'. *Isis*, 28(2), pp. 349–363.

Bowker, G.C. and Star, S.L., 2000. *Sorting Things Out: Classification and its Consequences*. Boston, MA: MIT Press.

Bragadóttir, K., 2017. 'Íslenskar bækur erlendis: Bókasöfnun Willards Fiskes (1831–1904)'. PhD Dissertation. University of Iceland.

Breitenbach, J. and Fischer, D., 2013. 'Pharma-Marketing: Sozioökonomische Trends bestimmen die Zukunft'. In Fischer, D. and Breitenbach, J. (eds).: *Die Pharmaindustrie: Einblick-Durchblick-Perspektiven*. Berlin: Springer, pp. 269–290.

Briquet, P., 1853. *Traité thérapeutique de quinquina et de ses préparations*. Paris: Victor Masson.

Britton, W.E., 1970. 'Some effects of science and technology upon our language'. *College Composition and Communication*, 21(5), pp. 342–346.

Brodovsky, J., 1995. *La industria farmacéutica y farmoquímica mexicana en el marco regulatorio de los años noventa*. Santiago: CEPAL.

Brown, L., 1920. 'Discussion of the present position of vitamins in clinical medicine'. *The British Medical Journal*, 160, pp. 160–162.

Brumme, J. (ed.), 2001. *La historia de los lenguajes iberorrománicos de especialidad: La divulgación de la ciencia*. Madrid: Iberoamericana Editorial Vervuert.

Buchanan, A. 2021. 'Plant of the Month: Tree of Life'. *JStor Daily*. 12 May. https://daily.jstor.org/plant-of-the-month-tree-of-life/ [Access October 2023]

Buisan, J.P., 2019. 'El "específico homeopático": Legitimación comercial de la homeopatía en Barcelona (1902–1910)'. *História, Ciências, Saúde-Manguinhos*, 26, pp. 1337–1354.

Burr, S. and Leung, Y.L., 2014. 'Curare (d-Tubocurarine)'. In: Wexler, P. (ed.) Encyclopaedia of Toxicology. London: Academic Press.

Cadet de Gassicourt, L.C., 1829. *Posología o Descripción de las cantidades de los Medicamentos que se prescriben en las enfermedades*. Puebla: Imprenta del C. Pedro de la Rosa.

Cahlíková, L., Šafratová, M., Hošťálková, A. et al., 2020. 'Pharmacognosy and its role in the system of profile disciplines in pharmacy'. *Natural Product Communications*, 15(9), pp. 1–6.

Cahours, A., 1850. 'Recherches sur les huiles légères obtenues dans la distillation du bois'. *Compte rendus*, 30, pp. 319–332.

Calatayud, J. and González, A., 2003. 'History of the development and evolution of local anesthesia since the Coca Leaf'. *Anesthesiology*, 98, pp. 1503–1508.

Campbell, L., 2013. *Historical Linguistics: An Introduction*. 3rd edn. Cambridge, MA: MIT Press.

Cancino, R., 2017. 'Las amenazas ambientales contra el hábitat de las plantas medicinales: La medicina mapuche en Chile'. *Sociedad y Discurso*, 31, pp. 66–87.

Cañizares-Esguerra, J., 2006. *Nature, empire, and nation: Explorations of the history of science in the Iberian world*. Stanford, CA: Stanford University Press.

Carli, A., and Calaresu, E., 2007. "Language and Science." In *Handbook of Language and Communication: Diversity and Change*, edited by Marlis Hellinger and Anne Pauwels, 523–552. *Handbooks of Applied Linguistics (HAL)*, vol. 9.

Carpenter, K.J., 2012. 'The discovery of vitamin C'. *Annals of Nutrition and Metabolism*, 61(3), pp. 259–264. doi:10.1159/000343121.

Cartwright, A.C. and Armstrong, N.A., 2020. *A History of the Medicines We Take: From Ancient Times to Present Day*. Yorkshire: Pen and Sword History.

Castellanos, A., 1957. 'Que ha de ofrecer Vd. a este importante evento'. *Kuba: Revista de medicina tropical y parasitología*, 13, p. 99.

Cengizhan, L. and Tanış, A., 2010. 'Purification of the medical terms in Turkish'. *Procedia-Social and Behavioral Sciences*, 2(2), pp. 5326–5330.

Cervantes, J., 1825. *Tablas botanicas que, para el mas pronto y facil estudio de esta ciencia*. Puebla: Moreno Hermanos.

Chaganti, S.R., 2006. *Pharmaceutical marketing in India*. New Delhi: Excel Books India.

Chattopadhyay, M. and Sahana, P.K., 1923. 'A Tale of Discovery of Insulin'. In Pandit, K. (ed.): *Insulin Discovery: The First Centenary Celebration Symposium*. Calcutta: Department of Endocrionology.

Chemla, K. and Virbel, J. (eds.), 2015. *Texts, Textual Acts and the History of Science*. Chicago, IL: University of Chicago Press.

Chilisa, B., 2019. *Indigenous Research Methodologies*. Los Angeles: Sage Publications.

Clark, F., 2009. 'Read all about it: Science, translation, adaption, and confrontation in the Gazeta de literaturade México (1788–1795)'. In: Bleichmar, D., De Vos, P., Huffine, K. and Sheehan, K. (eds.) *Science in the Spanish and Portuguese Empires, 1500–1800*. Stanford: Stanford University Press, pp. 147–177.

Clavería, G., 2016. *De vacunar a dictaminar: La lexicografía académica decimonónica y el neologismo*. Madrid: Iberoamericana Vervuert.

Cohen, G. and Moliner, P., 2022. 'Analogic and symbolic dimensions in graphic representations associated with patient information leaflets for medicines'. *Visual Studies*, 37(5), pp. 491–501.

Comisión de Investigaciones Científicas de la Provincia de Buenos Aires, 1962. *Memoria*. Buenos Aires: CICBA.

Conklin, H.C., 1962. 'Lexicographical treatment of folk taxonomies'. *International Journal of American Linguistics*, 28(2), pp. 119–141.
Conklin, H.C., 1980. *Folk classification: A topically arranged bibliography of contemporary and background references through 1971*. Revised reprinting with author index.
Cornips, L., 2024. The semiotic repertoire of dairy cows. *Language in Society*, 53(5), pp. 1–25.
Cowen, D.E., 2014. 'Logistics'. In: Adey, P., Bissell, D., Hannam, K., Merriman, P. and Sheller, M. (eds.) *The Routledge Handbook of Mobilities*. London: Routledge, pp. 187–195.
Cowen, D.L. and Helfand, W.H., 1990. *Pharmacy: An Illustrated History*. New York, NY: Harry N. Abrams.
Crawford, M.J., 2014. 'An empire's extract: Chemical manipulations of cinchona bark in the eighteenth-century Spanish Atlantic world'. *Osiris*, 29(1), pp. 215–229.
Crawford, M.J. and Gabriel, J.M., 2019. *Drugs on the Page: Pharmacopoeias and Healing Knowledge in the Early Modern Atlantic World*. Pittsburgh, PA: University of Pittsburgh Press.
Crilly, J., 2007. 'The history of clozapine and its emergence in the US market: A review and analysis'. *History of Psychiatry*, 18(1), pp. 39–60.
Cruz León, A., 2008. 'Los saberes campesinos en la obra de José Antonio Alzate Ramírez'. *Revista de Geografía Agrícola*, 40, pp. 103–105.
Cuesta y Ckerner, J., 1868a. *Anatomía Patológica: Extracto de las asignaturas del primer año que tienen que estudiar los cirujanos de 3ª clase que aspiran al título de facultativos habilitados*. Madrid: Tomás Alonso.
Cuesta y Ckerner, J., 1868b. *Obstetricia: Extracto de las asignaturas que han de estudiar los médicos puros que aspiran al título de licenciados en medicina y cirugía por medio de estudios privados*. Madrid: Tomás Alonso.
Cuesta y Ckerner, J., 1870. *Manual Balneario de España: Obra escrita para la Biblioteca de los Profesores de Partido*. Madrid: Tomás Alonso.
Cuesta y Ckerner, J., 1878a. 'Vocabulario técnico de medicina y ciencias auxiliares'. *La Correspondencia Médica*, 13, pp. 46–47.
Cuesta y Ckerner, J., 1878b. *Vocabulario tecnológico de medicina, cirujía y ciencias auxiliares*. Madrid: Gregorio Juste.
Cuesta y Ckerner, J., 1883. *Vocabulario tecnológico de Medicina, Cirujía, Farmacia y Ciencias auxiliares, corregida, aumentada y enriquecida con más de 21000 voces por D. Eduardo Aragón Obejero y D. Favila Cuesta Armiño (1883–1892), Vol. 4*. Madrid: Gregorio Juste.
Curzan, A., 2014. *Fixing English: Prescriptivism and Language History*. Cambridge: Cambridge University Press.
Dabrowska, E., 2004. *Language, Mind and Brain*. Edinburgh: Edinburgh University Press.
Dagblaðið Vísir – DV, 1990. 28 May.
Dahlerup, V. (ed.), 1926. *Ordbog over det danske sprog*. Vol. 10. Copenhagen: Gyldendal.
Darwish, A., 2009. *Terminology and Translation: A Phonological-Semantic Approach to Arabic Terminology*. Melbourne: Writescope Publishers.

Dascal, M., 2002. 'Language as a cognitive technology'. *International Journal of Cognition and Technology*, 1(1), pp. 35–61.

Dauncey, E.A., Irving, J.T.W. and Allkin, R., 2019. 'A review of issues of nomenclature and taxonomy of *Hypericum perforatum* L. and Kew's Medicinal Plant Names Services'. *Journal of Pharmacy and Pharmacology*, 71(1), pp. 4–14.

Davenport-Hines, R.P.T. and Slinn, J., 1992. *Glaxo: A History to 1962*. Cambridge: Cambridge University Press.

Davies, M., 2016. 'Corpus del Español NOW'. *News on the Web*. [online] Available at: www.corpusdelespanol.org/now/ [Accessed December 2022].

de Alzate y Ramírez, J.A., 1788. *Gaceta de México*, 15 March.

de Enterría Sánchez, J.G., Ramírez Luengo, J.L., Carpi, E., Dalle Pezze, F., Navarro, C. and Gallardo San Salvador, N., 2016. 'Tres momentos fundamentales en la historia de las ideas para la formación del vocabulario científico en el siglo XVIII'. In: Garriga Escribano, C. and Pérez Pascual, J.I. (eds.), *Lengua de la Ciencia e Historiografía*, pp.83–98. A Coruña: Universidade da Coruña.De la Cal, L., 2015. 'La viagra para jóvenes de veinte a treinta'. *El Mundo*, 28 June. Available at: www.elmundo.es/cronica/2015/06/28/558d9ccfe2704e2f288b456c.html [Accessed December 2022].

de la Cruz, M., 1552. *Libellus de Medicinalibus Indorum Herbis* [Cruz-Badiano Codex]. Translated by Juan Badiano. Mexico.

de Mendoza Ibáñez, F.J.R., 2000. 'El modelo cognitivo idealizado de tamaño y la formación de aumentativos y diminutivos en español'. *Revista Española de Lingüística Aplicada*, 1, pp. 355–374.

De Miguel, R. and Gómez de la Cortina, J. (Marqués de Morante), 1867. *Nuevo diccionario latino-español etimológico*. Leipzig: F.A. Brockhaus.

De woningbouwvereniging, 1950. *Congres-1950 Nationale Woningraad*, 39–41.

Dear, P. (ed.), 2015. *The Literary Structure of Scientific Argument: Historical Studies*. Philadelphia, PA: University of Pennsylvania Press.

DECEL (Diccionario Etimológico Castellano en Línea), 2022. 'Curarizar'. [online] Available at: http://etimologias.dechile.net [Accessed February 2023].

Degaetano-Ortlieb, S. and Teich, E., 2022. 'Toward an optimal code for communication: The case of scientific English'. *Corpus Linguistics and Linguistic Theory*, 18(1), pp. 175–207. doi:10.1515/cllt-2018-0088.

Del Valle, J., 2009. 'Total Spanish: The politics of a pan-Hispanic grammar'. *PMLA, Publications of the Modern Language Association of America*, 124(3), pp. 880–886.

Department of Public Health (DPH). 2023. Commmonwealth of Massachusetts. www.mass.gov/orgs/department-of-public-health. [Access February 2024].

Derrida, J. 1972. *La Dissémination*. Paris: Editions du Seuil.

Diccionario del español de México, 2023. 'Curarizar'. [online] Available at: https://dem.colmex.mx/ver/curarizar [Accessed February 2023].

Ding, Y., Jin, M., Li, S. and Feng, D., 2012. 'Smart logistics based on the internet of things technology: An overview'. *International Journal of Logistics Research and Applications*, 24(4), pp. 323–345.

Direcció General d'Atenció i Prevenció de la Drogodependència, 1993. *Drogodependencias: manual didáctico para educadores (con actividades dirigidas a menores en edad de E.G.B.)*. alencia: Bancaixa.

Donohue, J., 2006. 'A history of drug advertising: The evolving roles of consumers and consumer protection'. *The Milbank Quarterly*, 84(4), pp. 659–699.

Drummond, J., 1920. 'The nomenclature of the so-called accessory food factors (vitamins)'. *Biochemical Journal*, 14, p. 660.

Duffin, J., 2019. *Stanley's Dream: The Medical Expedition to Easter Island.* Montreal: McGill-Queen's University Press.

Duffull, S.B.,Wright, D.F.B., Marra, C.A. and Anakin, M.G., 2018. 'A philosophical framework for pharmacy in the 21st century guided by ethical principles'. *Research in Social and Administrative Pharmacy*, 14(3), pp. 309–316. doi:10.1016/j.sapharm.2017.04.049.

Durzan, D.J., 2009. 'Arginine, scurvy and Cartier's "tree of life"'. *Journal of Ethnobiology and Ethnomedicine*, 5(5), pp. 1–16. doi:10.1186/1746-4269-5-5.

DV, 1983. 'Stjörnuspá', 22 January.

Dyck, E. and Charles, A., 2021. 'New social history of pharmacy and pharmaceuticals'. *Canadian Bulletin of Medical History*, 38(S1), pp. S1–S5.

Džuganová, B., 2019. 'Medical language: A unique linguistic phenomenon'. *Jahr – European Journal of Bioethics*, 10(1), pp. 129–145.

Egereva, S.A. and Mikhlina, E.E., 1929. 'Экспериментальное испытание противорадцтцческого действия Jecorol'a'. *Kazan Medical Journal*, 25(6), pp. 651–653.

Eggersdorfer, M., Laudert, D., Létinois, U., McClymont, T., Medlock, J., Netscher, T. and Bonrath, W., 2012. 'One hundred years of vitamins: A success story of the natural sciences'. *Angewandte Chemie International Edition*, 51(52), pp. 12960–12990. doi:10.1002/anie.201205886.

Einarsdóttir, B., 1984. *Úr ævi og starfi íslenskra kvenna* . Reykjavík: Bókrún.

Embætti landlæknis (Icelandic Directorate of Health), 2018. 'Haetta Vegan Misnotkunar Lyfja'. [online] Available at: https://island.is/s/landlaeknir/frett/Haetta-vegna-misnotkunar-lyfja [Accessed December 2022].

Enfield, N.J. and Sidnell, J., 2022. *Consequences of Language: From Primary to Enhanced Intersubjectivity*. Cambridge, MA: MIT Press.

Epstein, R.H., 2018. *The History of Hormones and How They Control Just About Everything*. New York, NY: W.W. Norton & Company.

Erlendsdóttir, E., 2015. 'Préstamos amerindios del Perú en las lenguas nórdicas'. *Tonos Digital*, 28, pp. 1–21.

Erlendsdóttir, E., Martinell Gifre, E. and Söhrman, I. (eds.), 2017. *De América a Europa: Denominaciones de alimentos americanos en lenguas europeas.* Madrid: Iberoamericana.

Evans, F.T., 1948. 'Recent advances in anæsthesia: Cantor lectures'. *Journal of the Royal Society of Arts*, 96(4779), pp. 703–741.

Everett, D., 2012. *Language: The Cultural Tool.* New York, NY: Vintage.

Faure, P., 2021. 'Du sildenafil citrate au Viagra® ou l'art délicat de nommer les médicaments'. *Terminology*, 20(1), pp. 74–91. doi:10.1075/term.20.1.04fau.

FDA (U.S. Food & Drug Administration), 2014. 'FDA Approves New Drug Treatment for Chronic Weight Management, First Since 2014'. [online] Available at: www.fda.gov/news-events/press-announcements/fda-approves-new-drug-treatment-chronic-weight-management-first-2014 [Accessed December 2022].

Feasby, W.R., 1958. 'The discovery of insulin'. *Journal of the History of Medicine and Allied Sciences*, 13(1), pp. 68–84.
Felitti, K. (2022, June 20). The Birth Control Pill and Family Planning. In Vergara, Á. (ed. In chief) *Oxford Research Encyclopedia of Latin American History*. Retrieved 9 Dec. 2024, from https://oxfordre.com/latinamericanhistory/view/10.1093/acrefore/9780199366439.001.0001/acrefore-9780199366439-e-1043.
Fernández Poyatos, M.D., 2011. 'La publicidad de salud en la prensa ilustrada de finales del siglo XIX'. *Questiones Publicitarias*, 1(16), pp. 108–124.
Fernández Rodríguez, R., 2023. 'Language, science and globalization in the eighteenth century'. *Berichte zur Wissenschaftsgeschichte*, 46(1), pp. 38–53.
Fields, A.M., Freiberg, C.S., Fickenscher, A. and Shelley, K.H., 2008. 'Patients and jargon: Are we speaking the same language?' *Journal of Clinical Anesthesia*, 20(5), pp. 343–346.
Figueroa Saavedra, M. and Berzaín Varela, G.R., 2023. 'Publicidad en salud y bilingüismo en Nueva España del siglo XVIII: El caso del doctor Joseph Bartolache'. *Historia (Santiago)*, 56(2), pp. 33–75.
Finsen, J., 1874. *Iagttagelser Angaaende Sygdomsforholdene i Island: Afhandling for den Medicinske Doktorgrad ved Københavns Universitet*. Copenhagen: C.A. Reitzel.
Fischer, J.G., 1844. Fischers Populære Naturlære, til Brug i Skoler og ved Selvunderviisning. Copenhagen: Frølund & Flinch. [online] Available at: www.kb.dk/e-mat/dod/130020224887_color.pdf [Accessed December 2022].
Fochler, M., 2016. 'Variants of epistemic capitalism: Knowledge production and the accumulation of worth in commercial biotechnology and the academic life sciences'. *Science, Technology and Human Values*, 41(5), pp. 922–948.
Folguera, P., 2022. 'De la transición política a la paridad'. In: Folguera, P. (ed.). *El Feminismo en España: Dos Siglos de Historia*. Madrid: Editorial Pablo Iglesias, pp. 163–202.
Foucault, M., 1966. *Les mots et les choses*. Paris: Éditions Gallimard.
Foucault, M., 1987 [1976]. *The History of Sexuality: An Introduction*. Translated by R. Hurley. New York, NY: Penguin Books.
Frake, C.O., 1977. 'Plying frames can be dangerous: Some reflections on methodology in cognitive anthropology'. *The Quarterly Newsletter of the Institute for Comparative Human Development*, 1(3), pp. 1–7.
Frake, C.O., 2012. 'The ethnographic study of cognitive systems'. In: Fishman, Y.J. (ed.), *Anonymous Readings in the Sociology of Language**. The Hague: Mouton, pp. 434–446.
Freysdóttir, J., Ómarsdóttir, S., Ingólfsdóttir, K., Vikingsson, A. and Ólafsdóttir, E.S., 2008. 'In vitro and in vivo immunomodulating effects of traditionally prepared extract and purified compounds from *Cetraria islandica*'. *International Immunopharmacology*, 8(3), pp. 423–430.
Friðriksson, H.K. (translator), 1880. *Eðlisfræði eptir Balfour Stewart: Með 48 Myndum*. Reykjavík: Hið Íslenska Bókmenntafélag.
Friedel, R., 1993. 'Some matters of substance'. In Lubar, S. D. and Kingery, W. D. (eds.): *History from Things: Essays on Material Culture*. Washington, DC and London: Smithsonian Institution Press, pp. 41–50.

Frühneuhochdeutsches Wörterbuch, 2023. [online] Available at: https://fwb-online.de/ [Accessed February 2023].

Funk, C., 1912. 'The etiology of the deficiency diseases: Beri-beri, polyneuritis in birds, epidemic dropsy, scurvy, experimental scurvy in animals, infantile scurvy, ship beri-beri, pellagra'. *The Journal of State Medicine*, 20(6), pp. 341–368. [online] Available at: www.jstor.org/stable/45202920 [Accessed December 2022].

Funk, C., 1916. 'Vitamins: A New Factor in Nutrition'. Presentation given at the New York Academy of Medicine meeting held on April 20, 1916. *Journal of the American Medical Association*, 66, pp. 1654–1656.

Gaceta Médica de México, 1932. 1 August.

Galán, G., 2010. '50 años de la píldora anticonceptiva'. *Revista Chilena de Obstetricia y Ginecología*, 75(4), pp. 217–220.

Gallegos, A.J., 1983. 'The Zoapatle I—A traditional remedy from Mexico emerges to modern times'. *Contraception*, 27(3), pp. 211–225.

Gänger, S., 2020. *A Singular Remedy: Cinchona Across the Atlantic World, 1751–1820.* Cambridge: Cambridge University Press.

Gangwal, A. and Gangwal, A., 2011. 'Naming of drug molecules and pharmaceutical brands'. *Journal of Current Pharmaceutical Research*, 7(1), pp. 1–5.

Gans, E.L., 1981. *The Origin of Language: A Formal Theory of Representation.* Los Angeles, CA: University of California Press.

García Cárdenas, C., 2019. '¿Salud o negocio? Productos milagrosos y su publicidad en la prensa gaditana de inicios del XIX'. *Revista Historia de la Comunicación*, 19, pp. 57–64.

García Delgado, J.L.G., Alonso, J.A. and Jiménez, J.C., 2013. *El español, lengua de comunicación científica.* Mexico City: Ariel.

García, J.M., 2016. 'Aumentativos y diminutivos'. In Gutiérrez-Rexach, J. (ed.).: *Enciclopedia de Lingüística Hispánica.* London: Routledge, pp. 416–422.

Garðarsdóttir, H., 2023. 'Eftirlendur, arðrán og umhverfi: Stigveldi baráttunnar gegn loftslagsbreytingum í Rómönsku Ameríku'. In *Fléttur.* VI, pp. 97–121. Reykjavik: Háskólaútgáfan.

Garriga Escribano, C., 2018. 'Las voces de la química'. In: Clavería Nadal, G. and Freixas Alàs, M. (eds.), *El diccionario de la Academia en el siglo XIX: la quinta edición (1817) al microscopio.* Madrid: Arco/Libros, pp. 313–337.

Garriga Escribano, C., 2019. 'Ciencia, técnica y diccionario en la primera mitad del siglo XIX: incubando revoluciones'. *ELUA Anexo*, 5, pp. 47–63.

Garriga Escribano, C. and Pérez Pascual, J.I., 2016. *Lengua de la ciencia e historiografía.* A Coruña: Universidade da Coruña.

Getz, F.M. (ed.), 2010. *Healing and Society in Medieval England: A Middle English Translation of the Pharmaceutical Writings of Gilbertus Anglicus.* Vol. 8. Madison, WI: University of Wisconsin Press.

Gibson, K., 1991. 'Tools, language and intelligence: Evolutionary implications'. Man, 26(2), pp. 255–264.

Gies, J.R., Mooney, M.T. and Laughlin, R.M., 2017. 'Local Anesthetics'. In: Fonseca, R. (ed.), *Oral and Maxillofacial Surgery-Inkling*, Vol. 3. St. Louis: Elsevier Health Sciences, pp. 176–190.

Giral, F., 1966. 'Chemistry and society, III: Phyto-chemistry and zoo-chemistry in Hispanic America'. *Impact of Science on Society*, XVI(4), pp. 227–276. Available at: https://unesdoc.unesco.org/ark:/48223/pf0000262084/PDF/013610engo.pdf.multi.nameddest=262084 [Accessed 23 March 2023].

Giroux, C., 2007. 'La contribution synergique des pratiques de publicité pharmaceutique à la transformation du rôle contemporain du médicament'. *Revue Internationale sur le Médicament*, 1, pp. 33–77.

Gíslason, K., 1851. *Dönsk orðabók með íslenzkum þýðingum*. Copenhagen: Bianco Luno.

Godínez Reséndiz, R. and Aceves Pastrana, P., 2014. 'El surgimiento de la industria farmacéutica en México (1917–1940)'. *Revista mexicana de ciencias farmacéuticas*, 45(2), pp. 55–68.

Goedecke, T., Morales, D.R., Pacurariu, A. and Kurz, X., 2018. 'Measuring the impact of medicines regulatory interventions–systematic review and methodological considerations'. *British Journal of Clinical Pharmacology*, 84(3), pp. 419–433.

Gómez, J.A., 2002. 'Reforzará Sanofri su expansión'. *El Universal*, 23 January. Available at: https://archivo.eluniversal.com.mx/finanzas/25375.html [Accessed 23 March 2023].

Gómez Saavedra, V.I., Carreño González, P.S., Escobar Peña, M.E., Irarrazabal Verdugo, A.D., Rubio Lagos, C.C. and Weinstein Oppenheimer, C.R., 2016. 'Historia, presente y proyecciones de la Farmacopea'. *Anales de la Real Academia Nacional de Farmacia*, 82(3), pp. 283–296.

Gonsalves, L. and Hindin, M.J., 2017. 'Pharmacy provision of sexual and reproductive health commodities to young people: A systematic literature review and synthesis of the evidence'. *Contraception*, 95(4), pp. 339–363.

Gonzalez, W.J., 2021. 'The relevance of language for scientific research'. In: Gonzalez, W. (ed.), *Language and Scientific Research*. Cham: Palgrave, pp. 1–37.

Gonzalez Rodriguez, M.F., 2022a. *Ontologies and Natures: Knowledge about Health in Visual Culture*. Lanham: Rowman and Littlefield.

Gonzalez Rodriguez, M.F., 2022b. *Indigeneity in Latin American Cinema*. New York: Bloomsbury.

Google Trends, 2023. [Accessed March 2023].

Gordh, T., Gordh, T.E., Lindqvist, K. and Warner, D.S., 2010. 'Lidocaine: The origin of a modern local anesthetic'. *The Journal of the American Society of Anesthesiologists*, 113(6), pp. 1433–1437.

Gordin, M., 2015. *Scientific Babel: How Science Was Done Before and After Global English*. Chicago, IL: University of Chicago Press.

Gordin, M.D., 2017. 'Introduction: Hegemonic languages and science'. *Isis*, 108(3), pp. 606–611.

Goury, G., 1900/1950. Advertisement of Atoquinol (allyl phenylcinchoninester), produced by Ciba Laboratories. Bibliothèques de Nancy, P-FG-ES-00795. Available at: https://galeries.limedia.fr/ark:/31124/dtcpwgcp1kvz46gv/ [Accessed 22 October 2024].

Govantes Morales, R., 2021. 'Sobre identidad y disciplina: artefactos de la memoria en la comunidad farmacéutica mexicana entresiglos XIX y XX'. In: Guevara Fefer, R. and García Murcia, M. (eds.), *Memoria y disciplinas: aproximaciones a la historia de las ciencias*. Mexico City: Bonilla Artegas, pp. 45–68.

Grabowski, Ł., 2015. 'Keywords and lexical bundles within English pharmaceutical discourse: A corpus-driven description'. *English for Specific Purposes*, 38, pp. 23–33. doi:10.1016/j.esp.2014.10.004.

Greene, J.A., 2014. *Generic: The Unbranding of Modern Medicine*. Baltimore, MD: John Hopkins University Press.

Grímsson, M., 1852. *Eðlisfræði eptir J.G. Fischer*. Copenhagen: S.L. Möller.

Grodzinsky, Y., Shapiro, L.P. and Swinney, D. (eds.), 2000. *Language and the Brain: Representation and Processing*. San Diego, CA: Academic Press.

Groenke, U., 1966. 'On standard, substandard, and slang in Icelandic'. *Scandinavian Studies*, 38(3), pp. 217–230.

Gröndal, B. (translator), 1879. *Efnafræði eftir H.E. Roscoe: Með 36 myndum*. Reykjavik: Hið íslenzka bókmenntafélag.

Gröndal, B., 1886. *Efnafræði*. Reykjavik: Prentsmiðja Sigm. Guðmundssonar.

Gröndal, B., 1923. *Dægradvöl (æfisaga mín)*. Edited by Þ. Edílonsson. Reykjavik: Bókaverzlun Á. Árnasonar.

Gruber, L., 1950. 'Preliminary report in the use of Xylocaine as a local anesthetic in dentistry'. *Journal of Dental Research*, 29(2), pp. 137–142.

Guðmundsdóttir, M., 2022. 'Mál á mannsævi. 70 ára þróun tilbrigða í framburði – einstaklingar og samfélag'. PhD Dissertation. University of Iceland. Available at: https://opinvisindi.is/bitstream/handle/20.500.11815/3419/Margrét%20Guðmundsdóttir%20-%20Doktorsritgerð%202022.pdf?sequence=1&isAllowed=y [Accessed 22 October 2024].

Guðmundsson, E., 2019. *Eðlisfræði Fischers, fyrsta eðlisfræðibókin sem kom út á íslensku*. Available at: https://uni.hi.is/einar/2019/11/25/edlisfraedi-fischers-fyrsta-edlisfraedibokin-sem-kom-ut-a-islensku/ [Accessed 22 October 2024].

Guðmundsson, Þ. and Einarsson, B., 1832. *Ármann á Alþíngi: eða almennur fundur Islendínga – Ársrit fyrir búhølda og bændafólk á Islandi*. Copenhaguen: C. Græbe.

Gunnarsdóttir, H., 2020. *Þættir úr sögu lyfjafræðinnar á Íslandi*. Reykjavik: Iðunn.

Gutiérrez Rodilla, B.M., 2010. 'Sobre lexicografía médica del renacimiento castellano: los vocabularios de Andrés Laguna y Bartolomé Hidalgo de Agüero'. *Revista de Lexicografía*, 16, pp. 59–74. https://doi.org/10.17979/rlex.2010.16.0.3805.

Gutiérrez Rodilla, B.M., 2012a. 'La obra lexicográfica de Manuel Hurtado de Mendoza: Sus diccionarios enciclopédicos de medicina'. *Asclepio. Revista de historia de la medicina y de la ciencia*, 64(2), pp. 467–490.

Gutiérrez Rodilla, B.M., 2012b. 'El vocabulario terminológico de Manuel Hurtado de Mendoza y el lenguaje médico de la España decimonónica'. *Revista de Filología Española*, 92(2), pp. 249–272.

Gutiérrez Rodilla, B.M., 2017. 'La preocupación por la lengua y su reflejo en la lexicografía: el caso de los vocabularios españoles de medicina en el siglo XIX y principios del XX'. *Moenia*, 23, pp. 583–602.

Gylfason, T. and Gylfason, T., 1985. 'New words for an old language'. *Diogenes*, 33(132), pp. 17–33.

Hallgrímsson, J., 1835. 'Um eðli og uppruna jarðarinnar'. *Fjölnir*, 1, Útlenzki og Almenni Flokkurinn, pp. 99–130. [https://timarit.is/page/2012171#page/n1/mode/2up] (https://timarit.is/page/2012171#page/n1/mode/2up).

Halliday, M.A.K. and Butt, D.G., 2019. 'Language and science, language in science, and linguistics as science'. In Thompson, G. et al. (eds.) *The Cambridge Handbook of Systemic Functional Linguistics*. Cambridge: Cambridge University Press, pp. 620–650.

Hannß, K., 2014. 'Reduplication strategies in Kallawaya'. In Danielsen, S., Hannß, K. and Zúñiga, F. (eds.) *Word Formation in South American Languages*. Amsterdam: John Benjamins, pp. 163–180.

Hannß, K. and Muyken, P., 2014. 'Reduplication in Andean languages'. In Gómez, G.G. and van der Voort, H. (eds.) *Reduplication in Indigenous Languages of South America*. Leiden: Brill, pp. 39–76.

Hanson, I.R. and Hingson, R.A., 1950. 'The use of Xylocaine, a new local anesthetic, in surgery, obstetrics and therapeutics: A preliminary report'. *Anesthesia & Analgesia*, 29(6), pp. 136–147.

Hardon, A., 2021. *Chemical Youth: Navigating Uncertainty in Search of the Good Life*. Cham: Springer Nature.

Haselow, A., 2012. 'Multilingual speakers in a monolingual society. From the history and present state of language planning in Iceland'. In Hüning, M., Vogl, U. & Moliner, O. (eds.) *Standard Languages and Multilingualism in European History*. Amsterdam: John Benjamins, pp. 127–151. Available at: https://neon.niederlandistik.fu-berlin.de/static/events/Haselow.pdf [Accessed 22 October 2024].

Hattori, A., 2014. 'History of pharmaceutical packaging in modern Japan. II– Package size of pharmaceuticals'. *Yakushigaku Zasshi*, 49(2), pp. 190–195.

Haugen, E., 1992. 'Review of 'Kjartan G. Ottosson, Íslensk málhreinsun: Sögulegt yfirlit [Icelandic language purification: Historical survey]'. *Language in Society*, 21(2), pp. 336–345.

Hauser, M.D., Chomsky, N. and Fitch, W.T., 2002. 'The faculty of language: What is it, who has it, and how did it evolve?' *Science*, 298(5598), pp. 1569–1579.

Hayes, E., Dua, R., Yeung, E. and Fan, K., 2017. 'Patient understanding of commonly used oral medicine terminology'. *British Dental Journal*, 223(11), pp. 842–845.

Héðinsdóttir, S.G., 2014. 'Sexism and Gender: Differences in Slang Terms in Icelandic and English as a Second Language' (Doctoral dissertation). Reykjavik: University of Iceland.

Helgadóttir, S., Svavarsdóttir, Á., Rögnvaldsson, E., Bjarnadóttir, K. and Loftsson, H., 2012. 'The Tagged Icelandic Corpus (MÍM)'. Proceedings of the Workshop on Language Technology for Normalisation of Less-Resourced Languages – SaLTMiL 8 – AfLaT2012. Istanbul, Turkey, pp. 67–72.

Hepler-Smith, E. and McEwen, L., 2019. 'A century of nomenclature for chemists and machines'. *Chemistry International,* 41(3), pp. 46–49.

Herman, B.L., 1992. *The Stolen House*. Charlottesville, VA: University Press of Virginia.

Hermann, S. and Zentner, M., 1933. 'Substances influencing the rise in nonprotein nitrogen induced in cats by vitasterin'. *Biochemische Zeitschrift,* 266, pp. 418–421.

Herrera, A., Gonzalez, F., Lago de la Vega, J.M., Perez, S. and Soriano, M., 1884. *Nueva farmacopea mexicana de la Sociedad Farmacéutica de México.* Mexico: Imprenta de Francisco Diaz de Leon.

Herrera, A.L., de Jauregui, M.F., Calderón, J.B. and Pérez, S., 1896. *Nueva farmacopea mexicana de la Sociedad Farmacéutica de México*, 3rd edition. Mexico: Oficina Tipográfica de la Secretaría de Fomento.

Hersch Martínez, P., 2001. 'La Farmacopea Nacional y el estatuto terapéutico de la flora en la biomedicina mexicana'. *Dynamis: Acta Hispanica ad Medicinae Scientiarumque Historiam Illustrandam*, 21, pp. 375–407.

Hess, A.F. and Unger, L.J., 1917. 'Prophylactic therapy for rickets in a Negro community'. *Journal of the American Medical Association*, 69(19), pp. 1583–1586.

Het dagblad: uitgave van de Nederlandsche Dagbladpers, 1946, 18 September.

Het Parool, 1955. 'Vaart'. 26 July.

Heubner, W., 1930. 'The after-effects of large doses of vitasterin'. *Klinische Wochenschrift*, 9, p. 775.

Historisches Lexikon der Schweiz (HLS), 2007. 'Chemische Industrie'. *Historisches Lexikon der Schweiz*, 22 February. Available at: https://hls-dhs-dss.ch/de/articles/014007/2007-02-22/ [Accessed 22 October 2024].

Ho, H.M.K., Xiong, Z., Wong, H.Y. and Buanz, A., 2022. 'The era of fake medicines: Investigating counterfeit medicinal products for erectile dysfunction disguised as herbal supplements'. *International Journal of Pharmaceutics*, 617, p. 121592.

Hobby, G., Clark, R. and Woywodt, A., 2022. 'A treasure from a barren island: The discovery of rapamycin'. *Clinical Kidney Journal*, 15(10), pp.1971–1972.

Hock, H.H. and Joseph, B.D., 2019. *Language History, Language Change, and Language Relationship*. Berlin: De Gruyter Mouton.

Hockett, C.F., 1959. '"Animal" languages and human language'. *Human Biology*, 31(1), pp. 32–39.

Hoffman, L., 2017. 'Pharmaceuticals and tourist spaces: Encountering the medicinal in Cozumel's linguistic landscape'. *ACME: An International Journal for Critical Geographies*, 16(1), pp. 59–88.

Hoffman, R., 2012. 'Under the surface of the chemical article'. In Kovac, J. and Weisberg, M. (eds.) *Roald Hoffman on the Philosophy, Art, and Science of Chemistry*. New York, NY: Oxford Academic, pp. 145–162.

Hoggarth, L., 2004. *Contributions to Cuzco Quechua Grammar*. Bonn: Shaker Verlag.

Holme, R. 2012. 'Cognitive linguistics and the second language classroom'. *TESOL Quarterly*, 46(1), pp. 6–29.

Hooper, R. and Klein, G., 1848. *Lexicon Medicum, Or, Medical Dictionary Containing an Explanation of the Terms in Anatomy, Human and Comparative, Botany, Chemistry, Forensic Medicine, Materia Medica, Obstetrics, Pharmacy, Physiology, Practice of Physic, Surgery, Toxicology, and the Different Branches of Natural Sciences Connected with Medicine, Together with a Variety of Information on All These Subjects*. London: Gilbert and Rivington.

Hopkins, F., 1929. 'The earlier history of vitamin research.' Nobel Lecture, December 11 1929. Nobel Prize Foundation. Available at: www.nobelprize.org/prizes/medicine/1929/hopkins/lecture/ [Accessed 22 October 2024].

Hudson, K., 1978. *The Jargon of the Professions*. New York: Springer.
Huguet-Termes, T., 2008. 'Islamic pharmacology and pharmacy in the Latin West: An approach to early pharmacopoeias'. *European Review*, 16(2), pp. 229–239.
Íðorðabankinn, 2023. Online website.
IFPMA, 2022. The Pharmaceutical Industry and Global Health. Available at: www.ifpma.org/publications/facts-and-figures-2022-the-pharmaceutical-industry-and-global-health/ [Accessed 22 October 2024].
Interpol, 2022. 'USD 11 million in illicit medicines seized in global INTERPOL operation'. Available at: www.interpol.int/en/News-and-Events/News/2022/USD-11-million-in-illicit-medicines-seized-in-global-INTERPOL-operation [Accessed 22 October 2024].
Ireland, M.E. and Mehl, M.R., 2014. 'Natural language use as a marker'. In Holtgraves, T. (ed.). *The Oxford Handbook of Language and Social Psychology*. Oxford: Oxford University Press, pp. 201–237.
Ísafold, 1925. 'Sólarljós og þorskalýsi', 14 May.
Ivanova, S., 2021. 'The system of terminology: History and science'. *SSRN*, 3850948.
Izquierdo, J.J., 1952. 'Origins and development of Mexican pharmacopoeiae'. *Bulletin of the History of Medicine*, 26(1), pp. 54–70.
Jack, D.B., 1997. 'One hundred years of aspirin'. *The Lancet,* 350(9075), pp. 437–439.
Jácome Roca, A., 2003. *Historia de los medicamentos*. Bogota: Academia Nacional de Medicina.
Jakovljevic, M., Liu, Y., Cerda, A., Simonyan, M., Correia, T., Mariita, R.M., Kumara, A.S., Garcia, L., Krstic, K., Osabohien, R. and Toan, T.K., 2021. 'The Global South political economy of health financing and spending landscape–history and presence'. *Journal of Medical Economics*, 24(sup1), pp. 25–33.
Janík, Z., 2016. English language as a cultural tool for intercultural communication. *CASALC Review*, 5(2), p. 39.
Jiang, W., 2000. 'The relationship between culture and language'. *ELT Journal*, 54(4), pp. 328–334.
Jóhannsson, M., 2011. 'Misuse of prescription medicines'. *Læknablaðið*, 10, p. 519.
Johnson, B., 1981. *Jacques Derrida: Dissemination*. London: The Athlone Press – The University of Chicago Press.
Johnson, E., 2016. 'Introduction: Glocal pharmaceuticalization'. In Johnson, E., Sjögren, E. and Åsberg, C. (eds.) *Glocal Pharma: International Brands and the Imagination of Local Masculinity*. London: Routledge, pp. 1–13.
Johnson, E. and Åsberg, C., 2016. 'Enrolling men, their doctors and partners: Individual and collective responses to erectile dysfunction'. In Johnson, E., Sjögren, E. and Åsberg, C. (eds.) *Glocal Pharma: International Brands and the Imagination of Local Masculinity*. London: Routledge, pp.75–88.
Johnson, E. and Åsberg, C., 2016. 'Conclusions: Glocal pharma and the Swedish Viagra man'. In Johnson, E., Sjögren, E. and Åsberg, C. (eds.) *Glocal Pharma: International Brands and the Imagination of Local Masculinity*. London: Routledge, pp. 99–108.

Johnson, E., Sjögren, E. and Åsberg, C. (eds.), 2016. *Glocal Pharma: International Brands and the Imagination of Local Masculinity*. London: Routledge.

Jones, A., 2012. Signage with Viagra Ads, Ensenada, Baja California Norte (Mexico). *Photograph*. Available at: https://commons.wikimedia.org/wiki/File:Signage_with_Viagra_Ads_-_Ensenada_-_Baja_California_Norte_-_Mexico_(6776103046).jpg [Accessed 22 October 2024].

Jones, O.F., 1964. 'Some Icelandic 'götumál' expressions'. *Scandinavian Studies*, 36(1), pp. 59–64.

Junod, S.W. and Marks, L., 2002. 'Women's trials: The approval of the first oral contraceptive pill in the United States and Great Britain'. *Journal of the History of Medicine and Allied Sciences*, 57(2), pp. 117–160.

Kahan, B.D., 2003. 'Discoverer of the treasure from a barren island: Suren Sehgal' (10 February 1932 to 21 January 2003). *Transplantation*, 76(3), pp. 623–624.

Kahl, O., 2022. *Sābūr ibn Sahl. The Small Dispensatory: Translated from the Arabic Together with a Study and Glossaries*. Leiden: Brill.

Katsanis, L.P., Pitta, D. and Morinville, A., 2012. 'Patient centricity: Lip service or genuine commitment? A qualitative examination of the pharmaceutical industry'. *International Journal of Pharmaceutical and Healthcare Marketing*, 15(1), pp. 106–133.

Kessel, N., 2013. "Doriden von Ciba': sleeping pills, pharmaceutical marketing, and Thalidomide, 1955–1963'. *History and Technology*, 29(2), pp. 153–168.

Kim, K.T., 2013. 'Macrolinguistically specified references to biomedicine in Korean oriental medical discourse: Forms and distributions'. 언어, 38(2), pp. 279–303.

Kim, J.Y., 2022. 'Adaptation of chemical terminology from English into Korean'. *English Language and Linguistics*, 28(3), pp. 39–57.

King, K. and Melzi, G., 2004. 'Intimacy, imitation and language learning: Spanish diminutives in mother-child conversation'. *First Language*, 24(2), pp. 241–261.

Klaus, P., 2009. 'Logistics research: A 50 years' march of ideas'. *Logistics Research*, 1(1), pp. 53–65.

Kovac, J. and Weisberg, M. (eds.), 2012. *Roald Hoffmann on the Philosophy, Art, and Science of Chemistry*. New York, NY: Oxford Academic.

Kragh, H., 2024. *The Names of Science: Terminology and Language in the History of the Natural Sciences*. New York: Oxford University Press.

Kramsch, C., 2014. 'Language and culture'. *AILA Review*, 27(1), pp. 30–55.

Kress, H., 2002. 'Um þýðingastarf og skáldskap Bjargar C. Þorláksson'. In Kristmundsdóttir, S.D. (ed.) *Björg: Ævisaga Bjargar C. Þorláksson*. Reykjavik: JPV, pp. 197–215.

Kripke, S., 1990. *Name and Necessity*. Oxford: Basil Blackwell.

Kristiansen, T., 2010. 'Conscious and subconscious attitudes towards English influence in the Nordic countries: Evidence for two levels of language ideology'. *International Journal of the Sociology of Language*, 204, pp. 59–95.

Kristinsson, A.P., 2004. *Orðmyndun: um leiðir til að auka íslenskan orðaforða*. Reykjavik: Íslensk málnefnd.

Kristinsson, A.P. and Hilmarsson-Dunn, A., 2013. 'The language situation in Iceland'. In Kaplan, R., Baldauf Jr., R. and Kamwangamalu, N. (eds.) *Language*

Planning in Europe: Cyprus, Iceland and Luxembourg. London: Routledge, pp. 100–169.

Kristinsson, A.P., 2014. 'Ideologies in Iceland: The protection of language forms'. In Hultgren, A.K., Gregersen, F. and Thøgersen, J. (eds.) *English in Nordic Universities: Ideologies and Practices*. Amsterdam: Benjamins, pp. 147–164.

Kristinsson, A.P., 2018. 'National language policy and planning in Iceland – aims and institutional activities'. In Stickel, G. (ed.) *National Language Institutions and National Languages*. Frankfurt: Peter Lang, pp. 243–249.

Kristinsson, A.P., 2022. 'Language policy research directions embedded in the sociolinguistic enterprise'. *Sociolinguistica*, 36(1–2), pp. 111–123.

Kristmundsdóttir, S.D., 2001. *Björg. Ævisaga Bjargar C. Þorláksson*. Reykjavik: JPV.

Kristmundsdóttir, S.D., 2011. 'Hver var Björg C. Þorláksson og hvert var framlag hennar til vísindanna?' *Vísindavefurinn*, 23 December. Available at: www.visindavefur.is/svar.php?id=61540# [Accessed 22 October 2024].

Kucharz, E.J., 2020. 'Medical eponyms from linguistic and historical points of view'. *Reumatologia*, 58(4), pp. 258–260. DOI: 10.5114/reum.2020.98440.

Küng, G., 2012. *Ontology and the Logistic Analysis of Language: An Enquiry into the Contemporary Views on Universals*. Cham: Springer.

Kvaran, G., 2004a. 'English influence on the Icelandic lexicon'. *Nordic Journal of English Studies*, 3(2), pp. 143–152.

Kvaran, G., 2004b. 'Hvað er slangur gamalt fyrirbrigði?' *Vísindavefurinn*. Available at: www.visindavefur.is/svar.php?id=4464 [Accessed 22 October 2024].

Lafuente, A. and Giménez Toledo, E., 2020. 'La lengua de la ciencia y su (inaplazable) conexión con la sociedad'. *The Conversation*, 15 June.

Lafuente, A. and Valverde, N., 2004. 'Linnean botany and Spanish imperial biopolitics'. In Schiebinger, L. and Swan, C. (eds.) *Colonial Botany: Science, Commerce, and Politics in the Early Modern World*. Philadelphia: University of Pennsylvania Press, pp.134–147.

Landlæknir, 1933. *Heilbrigðisskýrslur (Public Health in Iceland): Samdar af landlækni eftir skýrslum héraðslækna og öðrum heimildum*. Reykjavik: Ríkisprentsmiðjan Gutenberg.

Laszlo, P., 2010. 'Être un scientifique, c'est apprendre à traduire la parole des choses'. *Hermès, La Revue*, 56(1), pp. 113–120.

Læknablaðið, 1930. 'Sólarljós og þorskalýsi', 1(1), p.17. Available at: https://timarit.is/page/5864649#page/n21/mode/2up [Accessed 22 October 2024].

LeBlanc, T.W., Hesson, A., Williams, A., Feudtner, C., Holmes-Rovner, M., Williamson, L.D. and Ubel, P.A., 2014. 'Patient understanding of medical jargon: A survey study of US medical students'. *Patient Education and Counseling*, 95(2), pp. 238–242.

Ledley, F.D., McCoy, S.S., Vaughan, G. and Cleary, E.G., 2020. 'Profitability of large pharmaceutical companies compared with other large public companies'. *Jama*, 323(9), pp. 834–843.

Lee, M.R., 2005. 'Curare: The South American arrow poison'. *Journal of the Royal College of Physicians of Edinburgh*, 35(1), pp. 83–92.

Lee, J.W., 2022. 'Locating Translingualism'. *Key Topics in Applied Linguistics*. Cambridge: Cambridge University Press. DOI: 10.1017/9781009105361.007.

Leikind, M.C., 1948. 'Banting's Miracle, The Story of the Discoverer of Insulin by Seale Harris, Elliot P. Joslin; Sir Frederick Banting by Lloyd Stevenson'. *Bulletin of the History of Medicine*, 22(1), pp. 109–111.

Leonard, S.P., 2011. 'Relative linguistic homogeneity in a new society: The case of Iceland'. *Language in Society*, 40(2), pp. 169–186.

Leonard, S.P. and Árnason, K., 2011. 'Language ideology and standardisation in Iceland.' In Kristiansen, T. and Coupland, N. (eds.) *Standard Languages and Language Standards in a Changing Europe*. Oslo: Novus Press, pp. 91–96.

Leonti, M., Casu, L., Sanna, F. and Bonsignore, L., 2009. 'A comparison of medicinal plant use in Sardinia and Sicily—De Materia Medica revisited?'. *Journal of Ethnopharmacology*, 121(2), pp. 255–267.

Levey, M., 1962. 'Mediaeval Arabic bookmaking and its relation to early chemistry and pharmacology'. *Transactions of the American Philosophical Society*, 52(4), pp. 1–79.

Levey, M., 1973. *Early Arabic Pharmacology: An Introduction Based on Ancient and Medieval Sources*. Leiden: Brill Archive.

Lévi-Strauss, C., 1966. *The Savage Mind*. Chicago: University of Chicago Press.

Li, B., 2022. 'Transadvertising of Western pharmaceutical products in *The Chinese Mail* (1895–1910)'. *Perspectives*, 30(2), pp. 343–359.

Li, H., Jiang, H. and Liu, J., 2017. 'Traditional Chinese medical therapy for erectile dysfunction'. *Translational Andrology and Urology*, 6(2), pp. 192–198

Liberman, A., 1988. 'Book review: 'Orðabók um slangur, slettur, bannorð og annað utangarðsmál'. *Scandinavian Studies*, 60(1), pp. 87–89.

Liberman, A., 2019. 'An etymological dog kennel, or dog eat dog: Icelandic setja upp við dogg, Engl. to lie doggo, Engl. dog, and Engl. it's raining cats and dogs.' *Scandinavistica Vilnensis*, 14, pp. 201–213.

Lind, J., 1753. *A Treatise of the Scurvy in Three Parts*. London: A. Millar. Digitalized Version.

Ljósmæðrablaðið, 1930. Sólarljós og þorskalýsi, 8(1), p.14. Available at: https://timarit.is/page/5433538?iabr=on#page/n15/mode/2up [Accessed 22 October 2024].

Lobo, F., 1977. 'La industria farmacéutica. Materiales para el estudio del crecimiento del capitalismo español en los años sesenta. Análisis económico de un sector industrial modelo'. Doctoral Dissertation Universidad Complutense, Madrid.

Lockhart, F.B., 1950. 'At the pitcher'. *The Tatler and Bystander*, 180.

Löfgren, N., 1948. 'Studies on Local Anesthetics. Xylocaine a New Synthetic Drug'. Doctoral Dissertation. Stockholm College. Stockholm: Hæggströms boktr.

Lögberg, 1927. 'Breyttir lífnaðarhættir. Lögberg', 26 May.

Lonati, E., 2022. 'New plants & new names: Botanical terminology in late modern English lexicography'. In Dattero, A. (ed) *Il Bosco: Biodiversità, Diritti e Culture dal Medioevo al Nostro Tempo*. Rome: Viella, pp. 235–255.

López-Muñoz, F. and Alamo, C., 2009. 'Monoaminergic neurotransmission: The history of the discovery of antidepressants from the 1950s until today'. *Current Pharmaceutical Design*, 15(14), pp. 1563–1586.

Ludwig, D., 2017. 'The objectivity of local knowledge. Lessons from ethnobiology'. *Synthese*, *194*(12), pp. 4705–4720.

Luján-García, C., 2020. 'The use of English initialisms and abbreviations in the field of pharmaceutical business communication in Spanish'. *Terminology*, 26(1), pp. 60–81.

Lyons, A.B., 1887. *The Pharmaceutical Era,* Vol. I. Detroit: D.O. Haynes & Company.

MacLeod, J.J.R., 1978. 'History of the researches leading to the discovery of insulin'. *Bulletin of the History of Medicine*, 52(3), p. 295.

MacNalty, A.S., 1964. 'History of diabetes'. *British Medical Journal*, 2(5401), p. 112.

Makowska-Wąs, J., Rzepiela, A. and Podolak, I., 2022. 'Florian Sawiczewski (1797–1876), founder of the pharmacognostic collection in Kraków'. *Journal of the History of Collections*, 34(2), pp. 275–286.

Malaver, I., 2018. 'Funciones del diminutivo en el español venezolano'. *Cuadernos de Lingüística de El Colegio de México*, 5(2), pp.5–44.

Maltz, A., 1992. *The Role of Language in the Discovery and Acceptance of Vitamins*. Chicago: Urbana University of Illinois.

Mannheim,B.,1998.'"Time,notthesyllables,mustbecounted":Quechuaparallelism, word meaning, and cultural analysis'. *Michigan Discussions in Anthropology*, 13(1), pp. 238–281.

Mao, Y. and Fung, K.W., 2020. 'Use of word and graph embedding to measure semantic relatedness between Unified Medical Language System concepts'. *Journal of the American Medical Informatics Association*, 27(10), pp. 1538–1546.

Marko, V., 2020. *From Aspirin to Viagra: Stories of the Drugs that Changed the World*. Cham: Springer.

Martin, P.M. and Martin-Granel, E., 2006. '2,500-year evolution of the term epidemic'. *Emerging Infectious Diseases*, 12(6), pp. 976–980. DOI: 10.3201/eid1206.051263.

Martínez Solís, S., Aceves Pastrana, P. and Morales Cosme, A., 2007. 'Una nueva identidad para los farmacéuticos: La Sociedad Farmacéutica Mexicana en el cambio de siglo (1890–1919)'. *Dynamis: Acta Hispanica ad Medicinae Scientiarumque Historiam Illustrandam*, 27, pp. 263–285.

Mas-Bagà, M., 2023. '¿Qué argot utilizan los drogodependientes entre ellos?'. CAT Barcelona. Available at: www.cat-barcelona.com/faqs/view/que-argot-utilizan-los-drogodependientes-entre-ellos/ [Accessed 22 October 2024].

Masliyah, S.H., 1984. 'Incorporation of scientific and technological terminology in the Arab World and Israel: A comparative study'. In *Deseret Language and Linguistic Society Symposium*, Vol. 10, No. 1, p. 16.

Matthews, B.R. and Cartwright, A.C. (eds.), 1991. *Pharmaceutical Product Licensing: Requirements for Europe*. London: CRC Press. DOI: 10.3109/9780203984000.

Mattiello, E., 2008. *An Introduction to English Slang: A Description of its Morphology, Semantics and Sociology*. Vol. 2. Milan: Polimetrica-International Scientific Publisher.

McCarthy, J., Cassidy, I., Graham, M.M. and Tuohy, D., 2013. 'Conversations through barriers of language and interpretation'. *British Journal of Nursing*, 22(6), pp. 335–339.

McTavish, J.R., 1999. 'What did Bayer do before aspirin? Early pharmaceutical marketing practices in America'. *Pharmacy in History*, 41(1), pp. 3–15.

Medina-Córdoba, M., Cadavid, S., Pérez-Acosta, A.M. and Amaya-Giraldo, V., 2012. 'Factors that facilitate and hinder the comprehension of patient information leaflets (PILs): A brief scoping review'. *Frontiers in Pharmacology*, 12, p. 740334.

Meijer, E., 2019. *Animal Languages: The Secret Conversations of the Living World*. London: Hachette.

Mendoza, G. and Herrera, A., 1866. 'El Yoloxóchitl'. *Gaceta médica de México: Órgano de la Sociedad Médica de México*, II, pp. 223–224.

Mikaili, P., Sharifi, M., Shayegh, J. and Sarahroodi, S., 2012. 'A review on pharmacognotic and pharmaceutical terms originated from Islamic sources'. *Journal of Basic Applied and Scientific Research*, 2(4), pp. 3235–3241.

Miller, K., 2023. Cara Ozempic, el inesperado efecto secundario del fármaco usado para adelgazar. *Men's Health*. Available at: www.menshealth.com/es/salud-bienestar/a42774606/cara-ozempic-efecto-secundario/ [Accessed 22 October 2024].

Millones Espinosa, M. and Hernández Maluenda, N., 2020. 'Las emociones de la mujer como estrategia publicitaria del campo farmacéutico a comienzos del siglo XX en Chile'. *Anagramas-Rumbos y Sentidos de la Comunicación*, 18(36), pp. 77–94.

Mitchell, M.E., 2015. 'Contentious cures: The rise and decline of lichens as European materia medica'. *Pharmacy in History*, 57(3–4), pp. 55–66.

Møllebæk, M., 2020. 'Rhetoric and science and technology studies'. In Bahri, P. (ed.) *Communicating About Risks and Safe Use of Medicines: Real Life and Applied Research*. Cham: Springer, pp. 267–284.

Mondémé, C., 2011. 'Animal as subject matter for social sciences: When linguistics addresses the issue of a dog's 'speakership''. In Gibas, P., Pauknerová, K. and Stella, M. (eds.) *Non-Humans in Social Science: Animals, Spaces, Things*, pp. 87–105, Červený Kostelec: Pavel Mervart.

Montero, M. and Rodríguez-Martín, N., 2019. 'Cambios sociales y comunicativos a través de la publicidad médica y farmacéutica en la prensa histórica de Bilbao, 1885-1936'. *Revista de Comunicación y Salud*, 9(2), pp.27–46.

Montero, R., 2023. Official Site. Available at: www.rosamontero.es [Accessed 22 October 2024].

Monterrubio, J., 2017. EllaOne brand morning after pill blister pack. HRA Pharma. *Photograph*.

Morales Cosme, A.D. and Viesca Treviño, C., 2016. 'La zarzaparrilla y el análisis de la materia médica en el siglo XIX mexicano'. *Eä*, 8(1), pp. 37–61.

Morales Sarabia, A. and Aceves Pastrana, P., 2011. 'Los materiales de la Real Expedición Botánica a la Nueva España, en el contexto del naturalismo mexicano de finales del siglo XIX.' *Llull: Revista de la Sociedad Española de Historia de las Ciencias y de las Técnicas*, 34(73), pp. 61–80.

Morgan, M.S., 2017. 'Narrative ordering and explanation'. *Studies in History and Philosophy of Science Part A*, 62, pp. 86–97.

Morgan, M.S., 2022. 'Narrative: A general purpose technology for science'. In Morgan, M., Hajek, K. and Berry, D. (eds.) *Narrative Science: Reasoning, Representing and Knowing Since 1800*. Cambridge: Cambridge University Press, pp. 3–30.

Morgunblaðið, 1925. 'Heilbrigðistíðindi', 29 September.

Morgunblaðið, 1926. 'Af Ísafirði', 7 August.

Muntaner i Mariano (ed.), 2008. *Cooperativa d'Apotecaris: La distribución farmacéutica al servicio de las Islas Baleares (1931–2008)*. Palma de Mallorca: Lleonard Muntaner.

Muñoz Armijo, L., 2018. 'De la resina a la queratina: La extensión semántica del patrón derivativo de las sustancias en-ina y su difusión en el español moderno'. In *Actas del X Congreso Internacional de Historia de la Lengua Española: Zaragoza, 7-11 de Septiembre de 2015*. Zaragoza: Institución Fernando el Católico, pp. 1241–1260.

Muralis, V., Šimkutė, V., Vičkačkaitė, V. and Žvirblys, A., 2023. 'Vaistiniai buteliukai (1919–1939 m.), rasti Užupio g. 15A, Vilniuje, jų turinio tyrimų rezultatai bei interpretacija'. *Archaeologia Lituana*, 24, pp.98–114.

Murphy, M.A., 2018. 'Early industrial roots of green chemistry and the history of the BHC Ibuprofen process invention and its quality connection'. *Foundations of Chemistry*, 20(2), pp.121–165.

New Oxford American Dictionary, 2022. Online version.

Nickon, A. and Silversmith, E., 1987. *Organic Chemistry: The Name Game. Modern Coined Terms and their Origins*. New York: Pergamon Press.

Nieto-Galan, A., 2016. *Science in the Public Sphere: A History of Lay Knowledge and Expertise*. London: Routledge.

Niiniluoto, I., 2014. 'Values in design sciences'. *Studies in History and Philosophy of Science Part A*, 46, pp. 11–15.

Norri, J., 1991. 'Notes on the origin and meaning of chemical terms in Middle English medical manuscripts'. *Neuphilologische Mitteilungen*, 92(2), pp. 215–236.

Núñez, Y.C., 2012. 'Para un baño de belleza... Salud y publicidad infantil desde la prensa en Cuba (1902-1959)'. *Pedagogía y Saberes*, 37, pp. 157–168.

O'Connell, R.L., Hartridge-Lambert, S.K., Din, N., St John, E.R., Hitchins, C. and Johnson, T., 2013. 'Patients' understanding of medical terminology used in the breast clinic'. *The Breast*, 22(5), pp. 836–838.

O'Hagan, L.A., 2021. 'Blinded by science? Constructing truth and authority in early twentieth-century Virol advertisements'. *History of Retailing and Consumption*, 7(2), pp. 162–192.

O'Neil, C.K. and Fetrow, C.W., 2007. Plants as Drugs' In Swarbrick, J. (ed.) Encyclopedia of Pharmaceutical Technology. Third edition, pp. 2901–2912, New York, NY: Informa Healthcare.

Oddsson, G., 1819. *Ordabók, sem inniheldr flest fágiæt, framandi og vandskilinn ord, er verda fyrir í dønskum bókum*. Copenhagen: Þorsteinn Rangel. Available at: https://archive.org/details/Ordabokseminnihe000152051v0GunnReyk/page/n3/mode/2up?view=theater [Accessed 22 October 2024].

Oguamanam, C., 2008. 'Local knowledge as trapped knowledge: Intellectual property, culture, power and politics'. *The Journal of World Intellectual Property*, 11(1), pp. 29–57.

Ólafsson úr Grunnavík, J., 1736–1811. *Íslensk-latnesk orðabók*. Hand-written manuscript. AM 433 1–2. Árnastofnun.

Online Etymology Dictionary, 2023. Available at: www.etymonline.com/word/dope [Accessed 22 October 2024].

Organización Panamericana de la Salud, 1949. *Boletín de la Oficina Sanitaria Panamericana*, Vol. 28.

Organon. 1966. Lindiol leaflet. Mexico City: Organon Mexicana S.A.

Ortiz Gómez, T. and Ignaciuk, A., 2018. 'Historia de la anticoncepción e historia oral, una revisión bibliográfica'. *Dynamis*, 38(2), pp. 293–302.

Ortiz Gómez, T. and Ignaciuk, A., 2019. *Anticoncepción, mujeres y género. La 'píldora' en España y Polonia (1960-1980)*. Madrid: Los Libros de la Catarata.

Óskarsson, V., 2022. 'Hugleiðing um slangur'. In: Bédi, B., Þorláksdóttir, H.J. and Friðriksdóttir, K. (eds.) *Perspectives on Language and Context: Festschrift in Honor of Birna Arnbjörnsdóttir*. Reykjavik: Stofnun Vigdísar Finnbogadóttur í erlendum tungumálum, pp. 160–165.

Ostrowski, P.F., 1986. 'Who discovered vitamins?'. *The Polish Review*, pp. 171–183.

Ottósson, K., 1990. *Íslensk málhreinsun: Sögulegt yfirlit*. Reykjavik: Íslensk málnefnd.

Oxford English Dictionary, 2022. Online version.

Palmer, L.J., 1946. 'Insulin in the treatment of diabetes mellitus'. *Journal of the American Medical Association*, 132(9), pp. 502–507.

Pálsson, S., 1789. *Um Kalkverkun af Jörðu og Steinum*. Copenhagen: Hið íslenzka Lærdómslistafélag.

Papí-Gálvez, N., Cambronero Saiz, B. and Ruiz-Cantero, M.T., 2007. 'El género como «nicho»: El caso de la publicidad farmacéutica'. *Feminismo/s*, 10, pp. 93–110.

Pariente, B. and Laharie, D., 2014. 'Why, when and how to de-escalate therapy in inflammatory bowel diseases'. *Alimentary Pharmacology & Therapeutics*, 40(4), pp. 338–353.

Parkhurst, C., 2007. 'A communication course for a linguistically diverse student population'. *American Journal of Pharmaceutical Education*, 71(2), p. 36.

Paskins, M., 2022. 'Thick and thin chemical narratives'. In Morgan, M., Hajek, K. and Berry, D. (eds.) *Narrative Science: Reasoning, Representing and Knowing Since 1800*. Cambridge: Cambridge University Press, pp. 267–286.

Pennycook, A., 2021. 'Reassembling linguistics: Semiotic and epistemic assemblages'. In Burkette, A. and Warhol, T. (eds.) *Crossing Borders, Making Connections: Interdisciplinary in Linguistics*. Boston/Berlin: De Gruyter, pp. 111–128.

Pennycook, A., 2024. *Language Assemblages*. Cambridge: Cambridge University Press.

Peset, J.L., 1987. 'La naturaleza como "símbolo" en la obra de José Antonio de Alzate'. *Asclepio*, 39(2), pp. 285–295.

Peters, P., Smith, A., Funk, Y. and Boyages, J., 2016. 'Language, terminology and the readability of online cancer information'. *Medical Humanities*, 42(1), pp. 36–41.
Pétursson, S., 1951. 'Um vítamin'. *Náttúrufræðingurinn*, 4, pp. 166–172.
Pieterse, A.H., Jager, N.A., Smets, E.M. and Henselmans, I., 2013. 'Lay understanding of common medical terminology in oncology'. *Psycho-Oncology*, 22(5), pp. 1186–1191.
Pinilla, J. and Lépinette, B. (eds.), 2015. *Traducción y difusión de la ciencia y la técnica en España (S. XVI-XIX)*. Valencia: Universitat de València, Institut Universitari de Llengües Aplicades Modernes (IULMA).
Pinker, S. and Jackendoff, R., 2005. 'The faculty of language: What's special about it?'. *Cognition*, 95(2), pp. 201–236.
Pinkerton, J., 1808. *A General Collection of the Best and Most Interesting Voyages and Travels in All Parts of the World*. London: Longman, Hurst, Rees, and Orme.
Pitman's Commercial Series, 1725. *The World's Commercial Products with Equivalents in French, German and Spanish*. London: Sir Isaac Pitman & Sons.
Pollio, A., De Natale, A., Appetiti, E., Aliotta, G. and Touwaide, A., 2008. 'Continuity and change in the Mediterranean medical tradition: Ruta spp. (Rutaceae) in Hippocratic medicine and present practices'. *Journal of Ethnopharmacology*, 116(3), pp. 469–482.
Þorbergsdóttir, Á., 2019. 'Frá nýyrðum til íðorða: Orðanefndir í 100 ár'. *Vélaverkfræðingafélagið*. Available at: www.vfi.is/media/frettir/fra-nyyrdum-til-idorda-Agusta.pdf [Accessed 22 October 2024].
Þorláksson, B.C., 1930. *Mataræði og Þjóðþrif*. Reykjavik: Ísafoldarprentsmiðja.
Þorlákson, B.C., 1933. *Daglegar Máltíðir*. Reykjavik: Ísafoldarprentsmiðja.
Portugal, E. and Nonnenmacher, S., 2024. 'Every word is a world: Loanword ideologies and linguistic purism in post-Soviet Armenia'. *Multilingua*, 43(3), pp. 331–364.
Þráinsdóttir, A.S. and Einarsdóttir, E.E., 2021. *Á sporbaug: Nýyrði Jónasar Hallgrímssonar*. Reykjavik: Sögur.
Premack, D., 1985. 'Gavagai! or the future history of the animal language controversy'. *Cognition*, 19(3), pp. 207–296.
PRESEEA Corpus, 2014 (in progress). Corpus del Proyecto para el estudio sociolingüístico del español de España y de América.'. Alcalá de Henares: Universidad de Alcalá. Available at: http://preseea.uah.es [Accessed 22 October 2024].
Price, C., 2017. 'The age of scurvy'. *Distillations*, 3(2), pp. 12–23.
Quianzon, C.C. and Cheikh, I., 2012. 'History of insulin'. *Journal of Community Hospital Internal Medicine Perspectives*, 2(2), pp. 1–3. DOI: 10.3402/jchimp.v2i2.18701.
Rabl, C.R.H., 1931. 'The question of the combination of white phosphorus and vitasterin'. *Archiv für experimentelle Pathologie und Pharmakologie*, 159, pp. 47–53.
Radick, G., 2007. *The Simian Tongue: The Long Debate about Animal Language*. Chicago: University of Chicago Press.

Ragnarsdóttir, G.Á., 1996. 'The Icelandic translator and purism'. *Perspectives*, 4(2), pp. 223–233. DOI: 10.1080/0907676X.1996.9961289.

Ragnarsson, J.Ó., 1979. *Næring og heilsa*. Reykjavik: Helgafell.

Rainsford, K.D., 2011. 'Fifty years since the discovery of ibuprofen'. *Inflammopharmacology*, 19(6), pp. 293–297.

Raleigh, W., 1596. *The Discoverie of the Large Rich, and Bewtifvl Empire of Gviana*. London: Robert Robinson.

Rama, 2017. Curare darts and quiver. Musée d'ethnographie de Genève, Item ETHAM 007525. *Photograph*. Available at: https://commons.wikimedia.org/wiki/File:Curare_darts_and_quiver-ETHAM_007525-IMG_9972-gradient.jpg [Accessed 22 October 2024].

Ramírez Sánchez, J., Rojas, L. and Anzola, F., 1969. 'Función hepática y anovulación iatrogénica'. *Revista Colombiana de Obstetricia y Ginecología*, 20(1), pp. 11–17.

Ramos, S.P., 2022. 'La traducción médica al servicio de la publicidad farmacéutica en el siglo XIX: El caso del éter acético, de la sal amoniaco y del ácido nítrico en tres notas paratextuales de María Antonia Gutiérrez Bueno y Ahoiz'. *Panace@*, 23(55), pp. 12–20.

Rando, D., 2009. 'The Cat's Meow: "Ulysses," animals, and the veterinary gaze'. James Joyce Quarterly, 46(3/4), pp. 529–543.

Rang, H.P., Dale, M.M., Ritter, J.M., Flower, R.J. and Henderson, G., 2012. *Rang & Dale's Pharmacology*. New York: Elsevier Health Sciences.

Ramsden, W., 1918. 'The vitamines'. *Journal of the Society of Chemical Industry*, 37, pp. 53–59.

Rayfield, D., 2023. 'The indigestible impact of English on the modern Georgian lexicon'. *Digital Kartvelology*, 2, pp. 14–28.

Real Academia de la Lengua Española, 1726-1739. *Diccionario de Autoridades*. Online edition: https://apps2.rae.es/DA.html [Accessed 22 October 2024].

Real Academia Española, 2013. *Corpus del Diccionario histórico de la lengua española* (CDH). Available at: https://apps.rae.es/CNDHE [Accessed 22 October 2024].

Reynoso, M.O., Díaz Flores, M., Islas Flores, H. and Schifter Aceves, L., 2017. 'Técnicas e instrumentos químico-farmacéuticos en México (1849–1925)'. *CIENCIA Ergo-Sum, Revista Científica Multidisciplinaria de Prospectiva*, 24(1), pp. 54–64.

Ringler, D., 1996. 'On the nature and origin of the earth (Um eðli og uppruna jarðarinnar)'. In Hallgrímsson, J. (ed.) *Jónas Hallgrímsson: Selected Poetry and Prose*. Madison: University of Wisconsin-Madison General Library System. Available at: https://digicoll.library.wisc.edu/Jonas/Edli/Edli.html [Accessed 22 October 2024].

Rit þess (konunglega) íslenzka Lærdómslistafélags IX, 1789. Copenhagen: Hið íslenzka Lærdómslistafélag.

Ritmálssafn Orðabókar Háskólans, 2023. *Stofnun Árna Magnússonar í íslenskum fræðum*. Available at: http://ritmalssafn.arnastofnun.is/ [Accessed 22 October 2024].

Robelo, C.A., 1904. *Diccionario de aztequismos: ó sea, catalogo de las palabras del idioma nahuatl, azteca ó mexicano : introducidas al idioma castellano bajo diversas formas : contribución al Diccionario Nacional*. Cuernavaca: Printed by the author.

Robelo, C.A., 1949. Diccionario de aztequismos, o sea, Jardín de las raíces aztecas. Palabras del idioma náhuatl, azteca o mexicano, introducidos al idioma castellano bajo diversas formas (contribución al diccionario nacional)
. Mexico: Ediciones Fuente Cultural. Available at: https://bibliotecavecina.files.wordpress.com/2015/06/robelo-cecilio-a-diccionario-de-aztequismos.pdf [Accessed 22 October 2024].

Roberts, F., 2004. 'Speaking to and for animals in a veterinary clinic: A practice for managing interpersonal interaction'. *Research on Language and Social Interaction*, 37(4), pp. 421–446.

Rodríguez González, F., 1985. 'Anglicismos en el argot de la droga'. *Atlantis*, 16(1–2), pp. 179–216.

Rodríguez-Ocaña, E., Ignaciuk, A. and Ortiz-Gómez, T., 2012. 'Ovulostáticos y anticonceptivos: El conocimiento médico sobre "la píldora" en España durante el franquismo y la transición democrática (1940–1979)'. *Dynamis*, 32(2), pp. 467–494.

Roe, D., 1953. 'Today's Women'. *The Key West Citizen*, 18 May.

Roldán-Vendrell, M. and Fernández-Domínguez, J., 2012. 'Emergent neologisms and lexical gaps in specialised languages'. *Terminology: International Journal of Theoretical and Applied Issues in Specialized Communication*, 18(1), pp. 9–26.

Romain, S., 2013. 'Pharmaceutical health care and Inuit language communications in Nunavut, Canada'. *International Journal of Circumpolar Health*, 72, p. 21409. DOI: 10.3402/ijch.v72i0.22295.

Roscoe, H.E., 1872. *Chemistry*. London: MacMillan and Co.

Rosenfeld, L., 2002. 'Insulin: Discovery and controversy'. *Clinical Chemistry*, 48(12), pp. 2270–2288.

Rögnvaldsson, E., 2022. *Alls konar íslenska: Hundrað þættir um íslenskt mál á 21. öld er safn fjölbreyttra þátta um íslensku*. Reykjavik: Forlagið.

Ruiz López, H., 1792. *Quinología o Tratado del Árbol de la Quina ó Cascarilla, con su Descripción y la de Otras Especies de Quinos Nuevamente Descubiertas en el Perú, del Modo de Beneficiarla, de su Elección, Comercio, Virtudes, y Extracto Elaborado con Cortezas Recientes*. Madrid: En la Oficina de la Viuda é hijo de Marín.

Sadegh-Zadeh, K., 2015. In *Handbook of Analytic Philosophy of Medicine*. Dordrecht: Springer.

Saladino García, S., 2017. 'José Antonio Alzate y Ramírez: Máxima figura de la cultura novohispana del siglo XVIII'. *La Colmena*, 21, pp. 83–90.

Saldaña, J.J. and Azuela, L.F., 1994. 'De amateurs a profesionales: Las sociedades científicas en México en el siglo XIX'. *Quipu: Revista Latinoamericana de Historia de las Ciencias y la Tecnología*, 11(2), pp. 135–171.

Sánchez, V., 2023. '¿Qué es la 'cara Ozempic' y cómo puede quitarse?' *El Periódico*. Available at: www.elperiodico.com/es/sanidad/20230301/cara-ozempic-que-es-como-quitar-dv-83473915 [Accessed 22 October 2024].

Sanchez Petrement, M., 2023. 'Historicizing psychedelics: Counterculture, renaissance, and the neoliberal matrix'. *Frontiers in Sociology*, 8, p. 1114523.

Sanchez-Ramos, J.R., 2020. 'The rise and fall of tobacco as a botanical medicine'. *Journal of Herbal Medicine*, 22, p. 100374. DOI: 10.1016/j.hermed.2020.100374.

Sanders, C.R., 1995. 'Killing with kindness: Veterinary euthanasia and the social construction of personhood'. *Sociological Forum*, 10(2), pp. 195–214.

Sanders, C.R., 2007. 'Mind, self, and human-animal joint action'. *Sociological Focus*, 40(3), pp. 320–336.

Santamaría-Pérez, I. and Marimón-Llorca, C., 2008. 'Entrela ciencia y el mercado: la divulgación médica a través de la publicidad escrita'. *Panace@*, 9(28), pp. 178–191.

Sarpur, 1997. 'A efni óháð spurningaskrám. Interview with Gunnjóna Sigrún Jensdóttir in 1985'. National Museum of Iceland, 7359/0. Available at: www.sarpur.is/Adfang.aspx?AdfangID=542928 [Accessed 22 October 2024].

Savage-Rumbaugh, E.S., 1990. 'Language as a cause-effect communication system'. *Philosophical Psychology*, 3(1), pp. 55–76.

Sæmundsson, B., 1923. 'Hugleiðingar um hagnýtingu á fiskúrgangi'. *Ægir*, 16(11), pp. 165–172.

Schifter Aceves, L., 2014. 'Las Farmacopeas Mexicanas en la construcción de la identidad nacional'. *Revista méxicana de ciencias farmacéuticas* , 45(2), pp. 43–54.

Schirripa, P., 2015. *La vita sociale dei farmaci: Produzione, circolazione, consumo degli oggetti materiali della cura*. Lecce: Argo.

Schmidt, J.A., 1811. *Lehrbuch der Materia Medica*. Vienna: Kupffer und Wimmer.

Schouten, P., Stepputat, F. and Bachmann, J., 2019. 'States of circulation: Logistics off the beaten path'. *Environment and Planning D: Society and Space*, 37(5), pp. 779–793.

Schrör, K., 2022. *Acetylsalicylic Acid*. Berlin-Boston: De Gruyter.

Schwartz, L.M. and Woloshin, S., 2019. 'Medical marketing in the United States, 1997–2016'. *Journal of the American Medical Association*, 321(1), pp. 80–96.

Searle, J.R., 1975. 'A taxonomy of illocutionary acts'. In Gunderson, K. (ed.) *Language, Mind, and Knowledge*. Minneapolis: University of Minnesota Press, pp. 344–369.

Sebba, M., 2023. 'Multimodal, multidimensional, multilingual: Informational and sociolinguistic hierarchies in multilingual product packaging'. In Włodarczyk, M., Tyrkkö, J. and Adamczy, E. (eds.) *Multilingualism from Manuscript to 3D*. New York: Routledge, pp. 227–255.

Sebeok, T.A., 1965. 'Animal communication: A communication network model for languages is applied to signaling behavior in animals'. *Science*, 147(3661), pp. 1006–1014.

Secretaría de Salud, 2014. *Programa de Acción Específico*. Mexico: D.R. Secretaría de Salud.

Segers, Y. and Van Molle, L. (eds.), 2022. *Agricultural Knowledge Networks in Rural Europe, 1700–2000*. Woodbridge: Boydell & Brewer.

SEGOB, 1977. *Diario Oficial de la Federación*, 4 October.

Semali, L.M. and Kincheloe, J.L. (eds.), 2002. *What is Indigenous Knowledge?: Voices from the Academy*. London: Routledge.

Semba, R.D., 2012. *The Vitamin A Story: Lifting the Shadow of Death*. Basel: Karger.

Senner, W., 2000. 'Magnús Stephensen: Precursor of the "Fjölnismenn" and Icelandic Romanticism'. *Scandinavian Studies*, 72(4), pp. 411–430. Available at: www.jstor.org/stable/40920255?seq=2 [Accessed 22 October 2024].

Senning, A., 2019. *The Etymology of Chemical Names: Tradition and Convenience vs. Rationality in Chemical Nomenclature*. Berlin: De Gruyter.

Sestino, A. and Amatulli, C., 2022. 'Branded vs. Generic drugs: The role of self-perceived seriousness of disease'. *International Journal of Pharmaceutical and Healthcare Marketing*, 16(1), pp. 22–39.

Shampo, M.A. and Kyle, R.A., 2006. 'John J.R. Macleod—Nobel prize for discovery of insulin'. *Mayo Clinic Proceedings*, 81(8), p. 1006.

Shanahan, T., Tran, T.P. and Taylor, E.C., 2019. 'Getting to know you: Social media personalization as a means of enhancing brand loyalty and perceived quality'. *Journal of Retailing and Consumer Services*, 47, pp. 57–65.

Shapin, S., 1984. 'Pump and circumstance: Robert Boyle's literary technology'. *Social Studies of Science*, 14(4), pp. 481–520.

Sherman, H.C., 1921. 'The vitamins'. *Physiological Reviews*, 1(4), pp. 598–630.

Sieck, G.C., 2021. 'Physiology in perspective: Insulin—100 years of physiological discovery'. *Physiology*, 36(3), pp. 132–133.

Sierra Ospina, N. and Tamayo Cano, L.H., 2016. 'La implementación del modelo adjunto de la enseñanza basada en contenidos en un laboratorio de química farmacéutica en una universidad pública colombiana'. *Lenguaje*, 44(2), pp. 341–369.

Sigurðsson, I., 2010. 'The Icelandic enlightenment as an extended phenomenon'. *Scandinavian Journal of History*, 35(4), pp. 371–390.

Simon-Maeda, A., 2008. 'Bigger is better: Masculinity and consumerist discourses in Viagra advertisements'. In Bou, P. and Santaemilia, J. (eds.) *Gender and Sexual Identities in Transition: International Perspectives*. Newcastle upon Tyne: Cambridge Scholars Publishing, pp. 24–40.

Simpson, S.S., Adams, N., Brugman, C.M. and Conners, T.J., 2018. 'Detecting novel and emerging drug terms using natural language processing: A social media corpus study'. *JMIR Public Health and Surveillance*, 4(1), e7726.

Singh, C., 2022. 'Science in the vernacular? Translation, terminology and lexicography in the Hindi scientific glossary (1906)'. *South Asian History and Culture*, 13(1), pp. 63–86.

Slow Food, 2024. Icelandic Fjallagrös. 'Ark of Taste'. Available at: www.fondazioneslowfood.com/en/ark-of-taste-slow-food/icelandic-fjallagros/ [Accessed 10 October 2024].

Smith, M., 2014. *Pharmaceutical Marketing in the 21st Century*. New York: CRC Press.

Snorrason, B., Brynjúlfsson, G., Sigurðsson, J., Eiríksson, M. and Finsen, V., 1850. *Ný félagsrit, gefin út af nokkrum Íslendingum: Tíunda ár*. Copenhagen: L. Klein.

Statista, 2023. Health, Pharma and Medtech. Available at: www.statista.com/statistics/263102/pharmaceutical-market-worldwide-revenue-since-2001/ [Accessed 22 October 2024].

Steffensen, J., 1984. 'Þættir úr sögu sjúkdóma á Íslandi'. *Læknablaðið*, 70, pp.181–189. Available at: https://timarit.is/page/5892203?iabr=on#page/n5/mode/2up/search/kirtlaveiki [Accessed 22 October 2024].

Steindórsson, S., 1981. *Íslenskir náttúrufræðingar 1600–1900*. Reykjavik: Bókaútgáfa Menningarsjóðs.

Stepanischev, E., 2020. Татмедторг, 'Bolknote'. Available at: https://bolknote.ru/all/tatmedtorg/ [Accessed August 2024].

Stephensen, M., 1818. *'Um Járn og Stál'. Margvíslegt Gaman og Alvara*. Reykjavik: Leirárgarðar- og Beitistaðaprent. Available at: https://timarit.is/page/970763#page/n0/mode/1up [Accessed 22 October 2024].

Stewart, F.E., 1894. 'Do the so-called elegant preparations of cod liver oil, from which the oil has been removed, possess therapeutic value?'. Proceedings of the American Pharmaceutical Association at the Annual Meeting, Vol. 51, pp. 287–292.

Stibbe, A., 2001. 'Language, power and the social construction of animals'. *Society & Animals*, 9(2), pp. 145–161.

Stohs, S. and Muhi-Eldeen, Z., 1990. 'The transition to more clinically oriented pharmacy education and the clinical practice of pharmacy'. *Journal of Clinical Pharmacy and Therapeutics*, 15(6), pp. 435–454.

Stong, M., 1987. 'Lexical overgeneration in Icelandic'. *Nordic Journal of Linguistics*, 10(2), pp. 181–205.

Stonier, P.D., Silva, H. and Lahon, H., 2007. 'Pharmaceutical medicine: History, global status, evolution and development'. *International Journal of Pharmaceutical Medicine*, 21, pp. 253–262.

Sufrin, C.B. and Ross, J.S., 2008. 'Pharmaceutical industry marketing: Understanding its impact on women's health'. *Obstetrical and Gynecological Survey*, 63(9), pp. 585–596.

Sugiyama, Y. and Seita, A., 2013. 'Kanehiro Takaki and the control of beriberi in the Japanese Navy'. *Journal of the Royal Society of Medicine*, 106(8), pp. 332–334. DOI: 10.1177/0141076813497889.

Sukalo, A., Catic, T., Skrbo, A., Zunic, L. and Masic, I., 2021. 'Mobile Pharmacies Throughout History'. *Materia Socio-Medica*, 33(2), p. 148.

Suran, M., 2023. 'As Ozempic's popularity soars, here's what to know about semaglutide and weight loss'. *Journal of the American Medical Association*, 329(19), pp. 1627–1629.

Surman, J., 2019. 'Terminology between chemistry and philology: A Polish interdisciplinary debate in 1900?'. *Centaurus*, 61(3), pp. 232–253.

Svavarsdóttir, Á., 2007. 'Talmál og málheildir-talmál og orðabækur'. *Orð og Tunga*, 9, pp. 25–50.

Terán Elizondo, M.I., 2004. 'José Antonio de Alzate y la crítica ilustrada y neoclásica en sus "Gacetas de Literatura de México"'. In Esquivel Estrada, N.H. (ed.) *Pensamiento Novohispano*, pp. 679–696, Zamora: El Colegio de Michoacán, Universidad Autónoma de Zacatecas..

Thiers, G. and McGinnis, L., 2011. 'Logistics systems modeling and simulation'. In *Proceedings of the 2011 Winter Simulation Conference (WSC)*. IEEE, pp. 1531–1541.

Thorakkattu, P., Khanashyam, A.C., Shah, K., Babu, K.S., Mundanat, A.S., Deliephan, A., Deokar, G.S., Santivarangkna, C. and Nirmal, N.P., 2022. 'Postbiotics: Current trends in food and pharmaceutical industry'. *Foods*, 11(19), p. 3094.

Thorsteinsson, S. and Sæmundsson, J., 1938. *Skrá yfir lyf úr forskriftasafni Lyfsalafélags íslands*. Reykjavik: n.p.

Tiefer, L., 2006. 'The Viagra phenomenon'. *Sexualities*, 9(3), pp. 273–294.

Tímarit Verkfræðingafélags Íslands, 1924. 'Advertisement section'. *Tímarit Verkfræðingafélags Íslands*, 1 March.

Toledo Union Journal, 1943. Willys Pot Shots, Toledo Union Journal, 26 March.

Tourainne, A., 1866. 'Expériences sur la tradescentia erecta (Yerba del pollo), (nom mexicain.)'. *Gaceta Médica de México: Órgano de la Sociedad Médica de México*, II, pp. 181–188.

Troike, R.C., 1983. 'Can language be tested?'. *Journal of Education*, 165(2), pp. 209–216.

Tsuda, Y. and Sano, T., 1996. 'Erythrina and related alkaloids'. In Cordell, G. (ed.) *The Alkaloids: Chemistry and Pharmacology*, Vol. 48, pp. 249–337. London: Academic Press.

Tylén, K., E. Weed, M. Wallentin, A. Roepstorff, and Frith, C. D., 2010. 'Language as a Tool for Interacting Minds'. *Mind & Language* 25(1), pp. 3–29.

Unda, R., 1986. *The Abuse of Antidiarrheal Agents and Antimicrobial Drugs in the Treatment of Infant Diarrhea in Mexico*. Washington: PRITECH.

United Nations, 2015. *Trends in Contraceptive Use*. New York: U.N.

Universitas Medica, 1965. *Universitas Medica*. Bogota: Universidad Javeriana.

UNODC (United Nations Office on Drugs and Crime), 2023. *World Drug Report 2022*. Vienna and New York: United Nations Publication.

Valdez Garza, D., 2022. *La "Gazeta de literatura de México" (1788–1795) como periódico-libro*. Mexico: UNAM.

Valiyeva, K., 2021. 'Comparative analysis of lexico-semantic characteristics of medical terms related to pharmacy in Azerbaijani and Arabic'. *Turan-Sam*, 13(52), pp. 439–445.

van der Geest, S., 2018. 'Pharmaceuticals'. In Callan, H. and Coleman, S. (eds.): *The International Encyclopedia of Anthropology*. London: Wiley, pp. 1–10.

Van Rompay, J. and Carter, J.E., 1990. 'Loperamide hydrochloride'. In Florey, K. (ed.): *Analytical Profiles of Drug Substances*, Vol. 19. London: Academic Press, pp. 341–365.

Vanoni, G. and Burgos, D., 2018. 'Caso de Enseñanza: Viagra en Ecuador'. *Revista Espacios*, 39(15), p. 32. Available at: http://w.revistaespacios.com/a18v39n15/18391532.html [Accessed 22 October 2024].

Varó, E.A., 2008. 'Language and health care: Food for thought'. *Panace@*, 9(28), pp. 161–168.

Vecchio, I., Tornali, C., Bragazzi, N.L. and Martini, M., 2018. 'The discovery of insulin: An important milestone in the history of medicine'. *Frontiers in Endocrinology*, 9, p. 613.

Vega y Ortega, R., 2016. 'Pharmaceutical studies during the Second Empire in the Gaceta Médica de México.' *História, Ciências, Saúde – Manguinhos*, 23(2), pp. 1–17. Available at: www.scielo.br/j/hcsm/a/wv5xSGcfvyqbyL9tBFtfpzF/?format=pdf&lang=en

Vikør, L., 2010. 'Language purism in the Nordic countries'. *International Journal of the Sociology of Language*, 204, pp. 9–30.

Vogl, U., 2012. 'Multilingualism in a standard language culture'. In Hünning, M., Vogl, U. and Moliner, O. (eds.) *Standard Languages and Multilingualism in European History*. Amsterdam: John Benjamins, pp. 1–44.

Voröld, 1919. 'Voröld', 22 July.

Vos, P. de, 2006. 'The science of spices: Empiricism and economic botany in the early Spanish Empire'. *Journal of World History*, 17(4), pp. 399–427.

Vos, P. de, 2007. 'Natural history and the pursuit of empire in eighteenth-century Spain'. *Eighteenth-Century Studies*, 40(2), pp. 209–239.

Vos, P. de, 2010. 'European materia medica in historical texts: Longevity of a tradition and implications for future use'. *Journal of Ethnopharmacology*, 132(1), pp. 28–47. DOI: 10.1016/j.jep.2010.05.035.

Wass, S., Kresge, N. and Bloomberg, 2024. 'Novo Nordisk's market value of $570 billion is now bigger than the entire Danish economy—creating a 'Nokia risk' for Denmark'. *Fortune*, 1 May. Available at: https://fortune.com/europe/2024/05/01/novo-nordisk-market-value-570-billion-bigger-than-danish-denmark-economy/ [Accessed 22 October 2024].

Watkins, E.S., 2009. 'From history of pharmacy to pharmaceutical history'. *Pharmacy in History*, 51(1), pp. 3–13.

Weeks, J., 2011. *The Languages of Sexuality*. London: Taylor & Francis Group.

Whelpton, M., Guðmundsdóttir Beck, Þ. and Jordan, F.M., 2021. 'The semantics and morphology of household container names in Icelandic and Dutch'. *Language Sciences*, 49, pp. 67–81.

Whyte, S.R., Van der Geest, S. and Hardon, A., 2002. *Social Lives of Medicines*. Cambridge: Cambridge University Press.

Wildsmith, J.A.W., 2011. 'Lidocaine: A more complex story than 'simple' chemistry suggests'. *Proceedings of the History of Anaesthesia Society*, 43, pp. 9–16.

Williams, D.M., Nawaz, A. and Evans, M., 2020. 'Drug therapy in obesity: A review of current and emerging treatments'. *Diabetes Therapy*, 11, pp. 1199–1216.

Winograd, T., 1972. 'Understanding natural language'. *Cognitive Psychology*, 3(1), pp. 1–191.

Wittgenstein, L., 1953. *Philosophische Untersuchungen*. Oxford: Blackwell.

Wolfgram, M., 2010. 'Truth claims and disputes in Ayurveda medical science'. *Journal of Linguistic Anthropology*, 20(1), pp. 149–165.

World Health Organization, 2017. Guidance on the Use of International Nonproprietary Names (INNs) for Pharmaceutical Substances. Geneva: World Health Organization. Available at: https://cdn.who.int/media/docs/default-source/international-nonproprietary-names-(inn)/who-pharm-s-nom-1570.pdf?sfvrsn=8358367e_4&download=true [Accessed 22 October 2024].

World Health Organization, 2024. Source: WHO Official Site. Available at: www.who.int [Accessed 10 October 2024].

Xu, M., Heiðmarsson, S., Ólafsdóttir, E.S., Buonfiglio, R., Kogej, T. and Ómarsdóttir, S., 2016. 'Secondary metabolites from cetrarioid lichens: Chemotaxonomy, biological activities and pharmaceutical potential'. *Phytomedicine*, 23(5), pp. 441–459.

Yap, K.Y., Chuang, X., Lee, A.J., Lee, R.Z., Lim, L., Lim, J.J. and Nimesha, R., 2009. 'Pharmaco-cybernetics as an interactive component of pharma-culture: Empowering drug knowledge through user-, experience-and activity-centered designs'. *International Journal of Computer Science Issues (IJCSI)*, 3(1), pp. 1–7.

Young, A., Tordoff, J., Leitch, S. and Smith, A., 2019. 'Patient-focused medicines information: General practitioners' and pharmacists' views on websites and leaflets.' *Health Education Journal*, 78(3), pp. 340–351.

Zadeh, L.A., 2004. 'Precipitated natural language (PNL)'. *AI Magazine*, 25(3), p. 74.

Index